NUCLEAR POWER

Fusion

James A. Mahaffey, Ph.D.

Facts On File
An Infobase Learning Company

For Luke Bennett Fletcher

FUSION

Facts On File, Inc.
An imprint of Infobase Learning
132 West 31st Street
New York NY 10001

Library of Congress Cataloging-in-Publication Data
Mahaffey, James A.
    Fusion / James A. Mahaffey.
        p. cm.—(Nuclear power)
    Includes bibliographical references and index.
    ISBN 978-0-8160-7653-6 (alk. paper)
    1. Fusion—Popular works. I. Title.
    QC303.M27 2012
    621.48'4—dc23   2011012505

Facts On File books are available at special discounts when purchased in bulk quantities for businesses, associations, institutions, or sales promotions. Please call our Special Sales Department in New York at (212) 967-8800 or (800) 322-8755.

You can find Facts On File on the World Wide Web at http://www.infobaselearning.com

Excerpts included herewith have been reprinted by permission of the copyright holders; the author has made every effort to contact copyright holders. The publishers will be glad to rectify, in future editions, any errors or omissions brought to their notice.

Text design by Annie O'Donnell
Composition by Julie Adams
Illustrations by Bobbi McCutcheon
Photo research by Suzanne M. Tibor
Cover printed by Yurchak Printing, Landisville, Pa.
Book printed and bound by Yurchak Printing, Landisville, Pa.
Date printed: January 2012
Printed in the United States of America

10 9 8 7 6 5 4 3 2 1

This book is printed on acid-free paper.

Contents

Nuclear Power is a multivolume set that explores the inner workings, history, science, global politics, future hopes, triumphs, and disasters of an industry that was, in a sense, born backward. Nuclear technology may be unique among the great technical achievements, in that its greatest moments of discovery and advancement were kept hidden from all except those most closely involved in the complex and sophisticated experimental work related to it. The public first became aware of nuclear energy at the end of World War II, when the United States brought the hostilities in the Pacific to an abrupt end by destroying two Japanese cities with atomic weapons. This was a practical demonstration of a newly developed source of intensely concentrated power. To have wiped out two cities with only two bombs was unique in human experience. The entire world was stunned by the implications, and the specter of nuclear annihilation has not entirely subsided in the 60 years since Hiroshima and Nagasaki.

The introduction of nuclear power was unusual in that it began with specialized explosives rather than small demonstrations of electrical-generating plants, for example. In any similar industry, this new, intriguing source of potential power would have been developed in academic and then industrial laboratories, first as a series of theories, then incremental experiments, graduating to small-scale demonstrations, and, finally, with financial support from some forward-looking industrial firms, an advantageous, alternate form of energy production having an established place in the industrial world. This was not the case for the nuclear industry. The relevant theories required too much effort in an area that was too risky for the usual industrial investment, and the full engagement and commitment of governments was necessary, with military implications for all developments. The future, which could be accurately predicted to involve nuclear power, arrived too soon, before humankind was convinced that renewable energy was needed. After many thousands of years of burning things as fuel, it was a hard habit to shake. Nuclear technology was never developed with public participation, and the atmosphere of secrecy and danger surrounding it eventually led to distrust and distortion. The nuclear power industry exists today, benefiting civilization with a respectable percentage of the total energy supply, despite the unusual lack of understanding and general knowledge among people who tap into it.

This set is designed to address the problems of public perception of nuclear power and to instill interest and arouse curiosity for this branch of technology. *The History of Nuclear Power,* the first volume in the set, explains how a full understanding of matter and energy developed as science emerged and developed. It was only logical that eventually an atomic theory of matter would emerge, and from that a nuclear theory of atoms would be elucidated. Once matter was understood, it was discovered that it could be destroyed and converted directly into energy. From there it was a downhill struggle to capture the energy and direct it to useful purposes.

Nuclear Accidents and Disasters, the second book in the set, concerns the long period of lessons learned in the emergent nuclear industry. It was a new way of doing things, and a great deal of learning by accident analysis was inevitable. These lessons were expensive but well learned, and the body of knowledge gained now results in one of the safest industries on Earth. *Radiation,* the third volume in the set, covers radiation, its long-term and short-term effects, and the ways that humankind is affected by and protected from it. One of the great public concerns about nuclear power is the collateral effect of radiation, and full knowledge of this will be essential for living in a world powered by nuclear means.

Nuclear Fission Reactors, the fourth book in this set, gives a detailed examination of a typical nuclear power plant of the type that now provides 20 percent of the electrical energy in the United States. *Fusion,* the fifth book, covers nuclear fusion, the power source of the universe. Fusion is often overlooked in discussions of nuclear power, but it has great potential as a long-term source of electrical energy. *The Future of Nuclear Power,* the final book in the set, surveys all that is possible in the world of nuclear technology, from spaceflights beyond the solar system to power systems that have the potential to light the Earth after the Sun has burned out.

At the Georgia Institute of Technology, I earned a bachelor of science degree in physics, a master of science, and a doctorate in nuclear engineering. I remained there for more than 30 years, gaining experience in scientific and engineering research in many fields of technology, including nuclear power. Sitting at the control console of a nuclear reactor, I have cold-started the fission process many times, run the reactor at power, and shut it down. Once, I stood atop a reactor core. I also stood on the bottom core plate of a reactor in construction, and on occasion I watched the eerie blue glow at the heart of a reactor running at full power. I did some time in a radiation suit, waved the Geiger counter probe, and spent many days and nights counting neutrons. As a student of nuclear technology, I bring a near-complete view of this, from theories to daily operation of a power plant. Notes and apparatus from my nuclear fusion research have been requested by and given to the National Museum of American History of the Smithsonian Institution. My friends, superiors, and competitors for research funds were people who served on the USS *Nautilus* nuclear submarine, those who assembled the early atomic bombs, and those who were there when nuclear power was born. I knew to listen to their tales.

The Nuclear Power set is written for those who are facing a growing world population with fewer resources and an increasingly fragile environment. A deep understanding of physics, mathematics, or the specialized vocabulary of nuclear technology is not necessary to read the books in this series and grasp what is going on in this important branch of science. It is hoped that you can understand the problems, meet the challenges, and be ready for the future with the information in these books. Each volume in the set includes an index, a chronology of important events, and a glossary of scientific terms. A list of books and Internet resources for further information provides the young reader with additional means to investigate every topic, as the study of nuclear technology expands to touch every aspect of the technical world.

Acknowledgments

I wish to thank Dr. Don S. Harmer, retired professor emeritus from the Georgia Institute of Technology school of physics, an old friend from the old school who not only taught me much of what I know in the field of nuclear physics, but did a thorough and constructive technical edit of the manuscript. I am also fortunate to know Dr. Douglas E. Wrege, a long-time friend and scholar with a Ph.D. in physics from the Georgia Institute of Technology, who is also responsible for a large percentage of my formal education. He did a further technical editing of the material. A particularly close, eagle-eyed edit was given the manuscript by my Ph.D. thesis adviser, Dr. Monte V. Davis, whose specific expertise in the topics covered in this work was extremely useful. Dr. Davis's wife, Nancy, gave me the advantage of her expertise, read the manuscript, and saved me from innumerable misplaced commas and hyphenations. Special credits are due Frank K. Darmstadt, my editor at Facts On File, Alexandra Simon, the copy editor, Suzie Tibor, the photography researcher, and Bobbi McCutcheon, the artist, who helped me at every step in making a beautiful book. The support and editing skills of my wife, Carolyn, were also essential. She held up the financial life of the household while I wrote and tried to make sure that everything was spelled correctly, all sentences were punctuated, and the narrative made sense to a nonscientist.

Introduction

The story of fusion power is every bit as exciting as the chronicle of the race for fission, but it has followed a completely different path. Fusion research and development have met their own set of unique difficulties and obstacles. The journey is still somewhere in the middle, between the initial discovery and practical application of fusion. The design of an operating fusion power plant may well be a problem to be conquered by a future generation, and the first steps outlined in this book may someday seem primitive indeed. The first fusion reactor connected into the power grid could be 50 or even 100 years in the future. Fusion power could be far enough away to be just in time to take over the power needs of civilization as uranium reserves are getting scarce. It is not too soon to start planning for how this fusion-powered world will operate. The next generation and the next after that have a great deal of scientific research to complete for this transition to occur.

The task of fusing two nuclei together to make energy is not particularly difficult. In August 1971, "How to Build a Machine to Produce Low-energy Protons and Deuterons," an article published in *Scientific American,* showed how a talented high school student could produce fusion in his or her basement. The problem is not making fusion. The problem is making sustainable fusion that actually makes more power than is used to initiate it. These problems have vexed the most brilliant scientists that the world has produced. It is a profound challenge.

At present, no one, no team of experts, and no international consortium seem able to make fusion work. Billions of dollars have been spent and hundreds of reactors have been built, but not a watt of usable power has been produced by a controlled fusion device. Unlike fission systems, precise prediction of fusion system behavior by mathematical means has proven difficult. Still, the advantages of this ultimate source of limitless power are too great to abandon. As energy problems of the world grow, work toward fusion power continues at a greater pace than ever before.

Fusion begins with the fundamental process of converting mass into energy, as is done on an astronomical scale, every minute of every day, by the Sun and all the stars in the universe. From the *big bang* of creation to the blackout death of wornout stars, fusion not only creates energy, but also creates everything from which the Earth and its inhabitants are made. A brief history of fusion research, beginning with the first tentative theories in the early 20th century, is covered in the second chapter. As was the case with fission, the first application of this new energy source was to build a bomb with it, and it was a rousing success. From that point, fusion research diverged from the straightforward path that fission seemed to take, and the quest continues.

The major line of inquiry then split into two lines, one pursuing magnetic confinement devices, as detailed in chapter 3, and the other going after inertial confinement devices, discussed in chapter 4. Both research programs experienced early encouraging results, but each has encountered obstacles in implementation on a larger scale.

These two major research paths are active and well funded, but they are not necessarily the only paths to successful controlled fusion. There are at least seven other ways to possibly make hydrogen fuse, and these exotic methods are outlined in chapter 5. Research continues in these areas, complicating the race for fusion power. The most promising research is still concentrated in magnetic and inertial confinement, and very large-scale prototype reactors are being built on these principles. Programs currently being funded or planned are discussed in chapter 6. *Fusion* concludes with an examination of the reality of fusion power and the magnitude of the challenge for future scientists and engineers.

Fusion expresses physical quantities in the traditional American engineering units, such as feet and pounds, with the international or SI units also provided for each measurement. An exception is temperature, which in this series is usually expressed as degrees Fahrenheit with parenthetical degrees Celsius. In the special case of a terrestrial fusion reaction, the temperatures are so high, in hundreds of millions of degrees, it does not matter what units are used. This unique fusion temperature expressed as Fahrenheit, Celsius, Kelvin, or even Rankine degrees is basically the same, given the extreme size of the number and the uncertainty involved in measuring it.

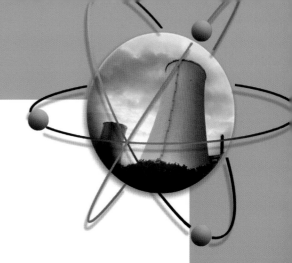

1 The Engine That Powers the Universe

In the final analysis, all energy has its origin in fusion. The Sun, which is a hydrogen-fusion reactor 864,328 miles (1,391,000 km) wide, has provided most of the energy on the planet Earth for the past 4.5 billion years. Sunlight warms the Earth and keeps it at a temperature that will sustain life in the otherwise cold void of outer space, but the Sun is responsible for much more. For thousands of years human beings have burned fuel, first wood and later coal, to heat homes, cook, light the darkness, pump water, move heavy loads, and form metals. The energy to do all this was provided by the Sun.

Wood, coal, and oil all have their origins in organic matter. Early in the development of life, plants evolved into systems that could collect sunlight and use the energy to bind two common elements, hydrogen and carbon, into building materials. By this process, a plant could grow tall, making it able to get above other plants and collect more sunlight. The sunlight that a plant prevents from hitting the ground by spreading a leaf to collect it is stored in the plant's structure. It is not lost energy; it is simply energy that has been converted to another form. It is stored and can be brought back at a later time. It could be released in a compost heap, in a fire, or buried underground as coal or oil. The energy can be derived as the heat of a fire almost immediately or can wait with supreme patience for millions of years.

When the energy is released, the chemical bonds holding together the carbon and hydrogen in the plant's structure are broken as the elements recombine with oxygen, forming two simple compounds, carbon dioxide and water. The heat of the Sun that was originally used to bind the hydrogen and carbon comes back in a fireplace on a cold winter's day, as if it had never left. Sunlight, in fact, puts about 93 watts per square foot (1,000 W/m^2) on the ground at noon on a sunny day, and there are presently many systems in use and in development to exploit this solar power. Power can be gathered directly from sunlight using photovoltaic cells or even using solar-heated steam boilers. In fact, all renewable energy derives indirectly from the Sun. Wind turbines, for example, use strong air currents blowing at near ground level to turn electrical generators, but the air currents are caused by the temperature differences in layers of air. Hot air tends to rise in the atmosphere, and cold air tends to fall, and shearing forces caused by this altitude exchange cause wind. The reason some air is warm is due, of course, only to the heat of the Sun.

In a similar fashion, hydropower is ultimately due to sunlight. Water on the land and seas is warmed by sunlight. It vaporizes in the heat and rises into the atmosphere. The vapor condenses into clouds in the cooler upper atmosphere, collects into water droplets, and falls as rain. The rain collects and runs in streams. Streams collect into rivers, and a river can be dammed, collecting water into a reservoir. By releasing backed-up water behind a dam, a water turbine spins a generator, and electricity is made. The motive power behind this scheme is the sunlight, lifting water, one molecule at a time, high into the air by heating it.

In the more recent development of civilization, more advanced forms of stored solar energy have been developed. Exploiting the extremely dense energy storage in the nucleus of certain special isotope variations of heavy elements has led to nuclear fission power. Uranium-235 can be made to fission, releasing vast amounts of stored energy in a self-sustaining reaction caused by neutron capture. Even this energy release strategy uses the power of fusion, but this was fusion not in the Sun, but in another star billions of years before the Sun was formed. As an older star ran out of fuel to fuse, it exploded into a *supernova* and managed to fuse some lesser elements into fissile uranium in the last seconds of its final destruction. The uranium and other constituent elements making up the planet Earth were the debris left over from this common stellar catastrophe. It was a fusion event in its most energetic form.

The electromagnetic spectrum spans a continuum from low-frequency radio waves to high-energy gamma rays. In the universe, the most common radiant energy occurs in an extremely narrow slice of this spectrum, which is in the subspectrum of visible light. Of all the radiant energy in the cosmos, all of which is made by the process of fusion, the most abundant radiation happens to occur at the visible frequency to which human eyes are the most sensitive, which is yellow-green light. Look up into the heavens on a clear night and marvel as countless numbers of fusion reactors each run continuously at full power, making primarily yellowish-green light.

THE PROCESS OF HYDROGEN FUSION

The most common material in the universe is simple hydrogen. A fundamental unit of hydrogen consists of a positively charged proton with a negatively charged electron orbiting it, resulting in a small particle of matter with its electrical charge cancelling out to zero. There are other, heavier isotopes of hydrogen having included neutrons at the center, but this simplest hydrogen is formally named hydrogen-1, indicating that it has an atomic mass of one, for its single proton nucleus. A hydrogen-1 atom weighs an insignificant 3.69×10^{-27} pounds (1.67×10^{-27} kg). From a statistical standpoint, everything in the cosmos is hydrogen-1, with some isolated exceptions.

Hydrogen-1 is the lightest isotope of the lightest existing element. The next heaviest element is helium, and its common isotope is helium-4, weighing almost four times as much as hydrogen-1. The important concept is that a helium-4 atom is almost as heavy as four hydrogen-1 atoms, because it is possible to assemble a helium-4 atom using four hydrogen-1 atoms. The weight of the assembled unit does not quite add up to that of the four component pieces, and that discrepancy, the mass deficit, is what warms the Earth.

In his famous theory of special relativity, Albert Einstein (1879–1955), a German theoretical physicist, presented a logical proof that mass and energy are equivalent. If mass is lost in a transaction such as the assembly of helium-4 from hydrogen-1 atoms, then it shows up as a conversion into energy. This principle is stated clearly in his equation

$$E = mc^2.$$

E is energy, m is mass, and c^2 is the conversion factor, the speed of light, or *celeritas,* squared. The conversion factor is quite large, about 9×10^{16}m/sec in SI units, indicating that very little mass converts to a great deal of energy.

The fusion of hydrogen into helium is the major source of power in stars the size of the Sun or smaller. The process, called the proton-proton reaction, is not quite as simple as bringing four hydrogen atoms in proximity to each other so that they can combine. It is a multistage chain reaction, requiring special conditions, and, as the most fundamental fusion process, it is worth a detailed analysis.

In fusion, the nuclei of two atoms fuse to form a larger nucleus. To accomplish this, they must be brought closely together, practically touching. This is normally impossible, as the nucleus of an atom is separated from all other atoms by the electron cloud surrounding it. In the case of hydrogen, there is only one electron, but it is still a very effective buffer against nuclear contact. If a hydrogen atom were the size of a football field, the proton at the center would be the size of a strawberry seed. It is made mostly of empty space. Take away the electron, and the positive charge on the proton is quite effective in keeping a like charged proton from getting anywhere near it.

For fusion to occur, the electron must be removed from the atom, leaving the nucleus naked and able to come into proximity to an adjacent nucleus. This is a state of matter, named the plasma state, that is a gaseous soup of disconnected electrons and atomic nuclei. As is the case of all matter, hydrogen can exist in four states, and the state of hydrogen is temperature-dependent. At its coldest, hydrogen is solid, and it melts at –434°F (–259°C), becoming a liquid. Liquid hydrogen boils at –423°F (–253°C), which is still quite cold. At room temperature, hydrogen is fully vaporized into a gas, but to become plasma and lose its orbiting electron, a hydrogen atom must be extremely hot. Hydrogen is in the plasma state at the center of the Sun, which is at 28,300,000°F (15,700,000°K).

Hydrogen without an orbiting electron is simply a proton, and a proton has a positive charge. Like charges tend to repel each other, so even though it has no electron shielding it from contact, a proton cannot normally get near another proton. The closer two protons come to each other, the stronger the repulsive force. However, there is a stronger force in effect. It is the nuclear binding force, which normally holds heavier elements together. This force is strong enough, at very close range, to overcome the powerful electrostatic repulsion and hold an atomic nucleus consisting of more than one proton together. If it is possible to overcome the electrostatic force and get two protons close enough together for the nuclear binding force to operate, then the two protons can stick together. The two positively charged particles do not have to be in touching distance for very long for this to happen, but still they must be brought close enough for the binding force to completely overcome the electrostatic force. This phenomenon of bringing two nuclei that repel each other close together is the main technical difficulty of

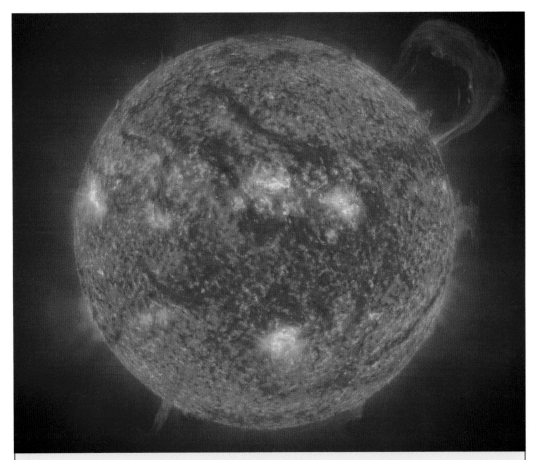

Extreme Ultraviolet Imaging Telescope (EIT) image of a handle-shaped prominence on the Sun, taken September 14, 1999 *(NASA Jet Propulsion Laboratory)*

PRE-FUSION THEORIES OF UNIVERSAL POWER

Before the advent of nuclear physics and quantum mechanics in the early 20th century, a puzzling problem for science was to explain the heat from the Sun. The Sun was obviously large and white-hot, with no apparent diminishing of size or temperature over human history, and the world's greatest minds had trouble with how this was accomplished. The great British scientist Lord Kelvin (1824–1907) teamed up with his distinguished German colleague Hermann von Helmholtz (1821–94) in the mid-19th century to explain the Sun's energy with a new study.

The Kelvin-Helmholtz theory held that any large ball of gas suspended in space will contract from mutual gravitational attraction of the gas particles. As the ball grows smaller, the gas, drawn to the center, converts the kinetic energy of falling by gravity into heat. Basically, the Sun gets its heat from friction, as the gas particles are stopped by crashing into adjacent particles. It made perfect sense, and it could be proven mathematically using well-established principles of Newtonian mechanics.

It was a sound theory, and it does explain how the Sun and stars became sufficiently hot and dense to initiate fusion at the core, but it had one serious problem. Given the known size of the Sun and the amount of hydrogen it contains, the Kelvin-Helmholtz theory gave the Sun a lifetime of 18 million years. At the time, there were fossils in rock formations found to be more than 300 million years old, and there

making terrestrial fusion power, but in a structure the size of a star it is inevitable.

What makes this process possible in the core of the Sun and stars is the high temperature of the plasma state and the fact that the pressure is extreme. The gaseous plasma at the center is pressed to a density of 9,400 pounds per cubic foot (150g/cc), or 150 times the density of water. The high temperature means that individual particles, or protons, are traveling in random directions at extremely high speed and are likely to crash into one another. Temperature is, in this case of plasma, simply a measure of the average speed of a proton, and at the Sun's core temperature a proton is traveling about 1.4 million miles per hour (630,000 m/sec). At that speed and density, fusion-causing collisions can occur.

were hints that Earth was, in fact, billions of years old. This was not possible, if the total lifetime of the Sun could be only 18 million years. It had been assumed that the Earth formed from interstellar debris and assumed orbit only after the Sun had formed and sunlight was necessary to sustain even the most primitive life-forms.

The British father of nuclear physics, Ernest, Lord Rutherford (1871–1937), documented an advanced solution in 1904, theorizing that the heat was produced by radioactive decay as an internal source of energy. This explanation was on the right track, but it was pure speculation. In 1920, Sir Arthur Eddington (1882–1944), a very capable British astrophysicist, proposed that the heat of the Sun and stars was produced by fusing four hydrogen atoms into one helium atom. The pressure and temperature at the center of the great ball of gas created the proper environment, and it explained spectroscopic analyses of sunlight, showing that the Sun was composed mostly of hydrogen, with a component of helium. The ultimate source of the heat could be explained by Albert Einstein's (1879–1955) recently revealed theory of mass-energy equivalence. The hydrogen obviously lost mass upon fusing, and it was this equivalent energy that was maintaining the fire of the Sun.

Eddington was correct, but it would take another 20 years of intense theoretical work by Hans Bethe (1906–2005), an American physicist born in Germany, to fully explain the mechanisms of hydrogen fusion. The Eddington limit, the maximum luminosity achievable by a star, is named in Arthur Eddington's honor.

A common misconception is that the purpose of building a terrestrial fusion power plant is to achieve the fusion conditions at the center of the Sun. The Sun produces power at the level of 3.85×10^{26} watts, meaning that it has 9.2×10^{37} proton-proton fusions per second. This seems a huge amount of power, but the Sun, as a moderately sized star, is inconceivably large. It weighs 2.19×10^{27} tons (1.99×10^{30} kg) and is 109 times as wide as the Earth. Fusion at the core actually occurs slowly, giving a power-production density of about 7.85 watts per cubic foot (277 watts/m^3), which approximately describes the metabolism of a reptile. An adult box turtle, awake and occupying 30 cubic inches (500 cc) of space, therefore generates as much heat as a similar volume of plasma at the center of the Sun. An active compost heap the size of the Sun would glow white from the heat buildup and illumine the Earth with warmth at the proper distance, 93 million miles (150 million km). The problem of achieving power production by fusion in an Earth-based reactor is therefore not to emulate the Sun. The problem is to create fusion that is several orders of magnitude greater than occurs in the Sun, which has a volume 1,300,000 times that of Earth. To do so requires greater core temperature and greater pressure in a structure that can occupy a moderately sized building.

Another way of visualizing the size of the Sun is to consider the time it takes for a photon to travel from its release in a proton-proton fusion at the center and travel to the surface, where it escapes as light. The photons are initially produced as high-energy gamma rays, but they collide with matter and are reflected off at random angles. In reflection, a photon is actually absorbed and then reemitted, at a slightly decreased energy. By the time a gamma ray has been reflected innumerable times as it makes its way to the Sun's surface, its energy has been reduced to visible light. The photon travel time, in a random path at the speed of light, is no shorter than 10,000 years and no longer than 170,000 years.

The *proton-proton chain reaction* begins with a collision, producing a highly unstable configuration of a proton stuck to a proton by nuclear binding force. One of the protons immediately decays into a neutron, and the combination becomes a *deuterium,* or hydrogen-2, nucleus. The proton decay ejects a positron, or antielectron, plus a neutrino and a modest 0.42 MeV of kinetic energy. The antielectron immediately finds an electron in the plasma soup, and the two annihilate one another. The energy of the matter destruction is carried off as two gamma rays, totaling 1.02 MeV of energy.

The deuterium then collides with another proton, or hydrogen-1 nucleus, and produces a light isotope of helium, or helium-3. The energy of this fusion is carried away as a single gamma ray, having an energy of 5.49 MeV. The total energy released from this sequence of events is 6.93 MeV.

From this point, there are three possible ways to make the final product, which is a common helium-4 nucleus. The production methods are chosen at random, with statistical preference given to the simplest scheme, which is designated pp I. In the pp I reaction (as shown in the figure on page 9), two helium-3 nuclei, left over from two examples of

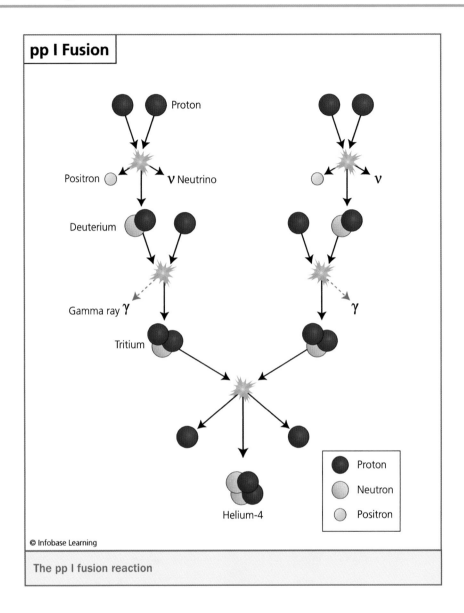

pp I Fusion

Proton

Positron ν Neutrino

ν

Deuterium

Gamma ray γ

γ

Tritium

	Proton
	Neutron
	Positron

Helium-4

© Infobase Learning

The pp I fusion reaction

the first reaction, collide and result in one helium-4 plus two freed protons, ejected with a kinetic energy of 12.86 MeV. This pp I branch of the creation of helium-4 from hydrogen-1 requires two complete creations of helium-3 nuclei, so the energy released going into the pp I reaction is 13.86 MeV. The total energy released by the pp I chain reaction is 26.72 MeV, and 86 percent of the proton-proton fusions are this mode.

To put this in perspective, a single fission of a uranium-235 nucleus releases more than 200 MeV of energy. Burning a single molecule of coal releases less than 0.000001 MeV of energy.

In a pp II reaction (see figure below), which occurs 14 percent of the time, a helium-3 nucleus from the first reaction plus a loose helium-4 from a previous climax reaction collide, producing a beryllium-7. The mass deficit of the transaction expresses itself as an energetic gamma ray. In an unusual move, the beryllium-7 then captures a free electron in the plasma soup, and this

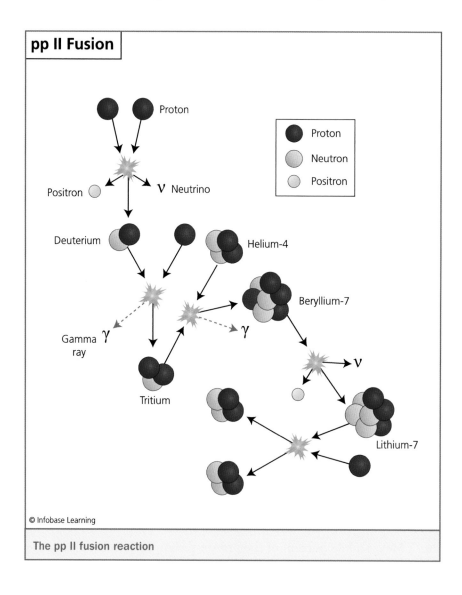

The pp II fusion reaction

converts one of the protons in the 4-proton beryllium nucleus to a neutron. The beryllium-7 is instantly converted to lithium-7, and the energy escapes

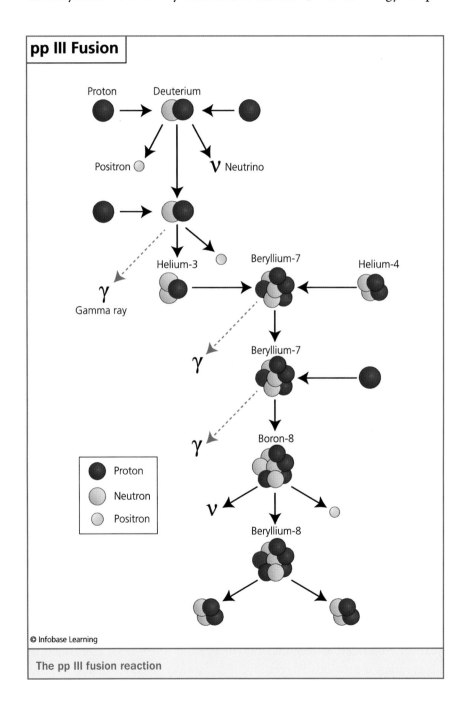

The pp III fusion reaction

as a neutrino. The neutrino energy is, unfortunately, unrecoverable, as the neutrino has very little chance of interacting with matter, even in a mass the size of the Sun. It simply escapes, unimpeded, and shoots off into space. The lithium-7 then captures another proton in the plasma, and it splits into a neat pair of helium-4 nuclei.

A more unlikely reaction, occurring only 10 percent of the time, is the pp III (shown in the figure on page 11). A helium-3 combines with a helium-4, becoming a beryllium-7 and releasing a gamma ray. The beryllium-7 then crashes into a proton, becomes boron-8, and releases another gamma ray. The boron-8 then experiences a beta-plus decay, in which one of its protons becomes a neutron, and it drops back to beryllium-8. A positron is released, plus a gamma ray. A very energetic neutrino carries away 14.06 MeV of the transactional energy with no possible recovery. The beryllium-8, being unstable, splits perfectly in half, producing two helium-4 nuclei.

There is another possible reaction, in which a helium-3 unites with a proton. The proton then immediately decays into a neutron, and the result is a helium-4 product with 18.8 MeV of energy. This theoretical reaction has never been observed.

A most interesting aspect of these reactions to produce helium from hydrogen is that they produce heavier elements as sidelines. Given the probabilistic character of nuclear events, some of these relay elements are bound to wind up as debris in the exhaust of the fusion process. This, plus other, more complicated reactions in stars of various sizes can explain how the building blocks of planets and other structures of the universe were created out of nothing but hydrogen, the simplest element.

BIG BANG NUCLEOSYNTHESIS: FUSION WHEN THE UNIVERSE WAS YOUNG

Nucleosynthesis is the process of building new atoms from previously existing components. The first instance of this phenomenon occurred about 14 billion years ago, in an event called the big bang, or the creation of the universe.

The creation event was subject only to speculation until scientific investigation began to uncover specific features, beginning in 1929. That year an American astronomer Edwin Hubble (1889–1953) confirmed by observation a previous suggestion, that faraway galaxies of stars are all

Edwin Hubble inside the workings of the 200-inch (508 cm) telescope at the Mount Palomar Observatory, California *(J. R. Eyerman/Time Life Pictures/Getty Images)*

moving away from our vantage point. Furthermore, the farther away a galaxy is, the faster it is traveling. This profound finding was made using the 100-inch (254-cm) reflecting telescope at Mount Wilson, in California. When Hubble began work at Mount Wilson in 1919, it was widely assumed that the Milky Way galaxy was the extent of the entire universe, and that fuzzy objects in space, called nebulae, were simply gas clouds. He was able to prove that while some nebulae were, in fact, gas clouds, others were distant galaxies and that our Milky Way galaxy was not the end of everything.

Hubble found evidence of Cepheid variable stars in these dim, fuzzy objects. Cepheid variable stars pulsate, blinking on and off with very regular periods from days to months. Moreover, Cepheids of specific periods always have the same luminosity and can be extremely bright. By finding a Cepheid-like pulsation in a nebula, it became clear to Hubble that he had discovered a distant, other galaxy. By measuring its luminosity, he could calculate the distance, given its pulsation period. He found that the largest nebula, Andromeda, is actually the Andromeda galaxy, and it is 2.5 million light-years away. It is the closest galaxy to the Milky Way. He went on to find 45 other galaxies and calculate their distances from Earth.

Another American astronomer Vestro Slipher (1875–1969) had been observing these same nebulae, or galaxies, using a spectrophotometer at the Lowell Observatory in Flagstaff, Arizona. Using the 24-inch (61-cm) Clark Refracting Telescope, he recorded the spectra of light from these nebulae, and he found them to be shifted slightly to the red end of the spectrum from the usual white light of familiar stars.

Using his observations of galactic distance and the spectra measurements of Slipher and others, Hubble was able to find a correspondence between the degree of red shift and the distance of a galaxy. The red shift of a galaxy was apparently due to the fact that it was moving rapidly away from Hubble's observing station. The speed of flight away tended to stretch out the light waves, shifting them toward the red end of the spectrum, by the well-established Doppler effect. The more the light from a galaxy was red-shifted, the faster it was fleeing. Hubble's finding was that the faster a galaxy was going, the farther away it was.

It was a logical extension of this finding that if everything observable in the larger universe is moving away, then everything must have been closer together at an earlier time. Monsignor Georges Lemaître (1894–1966), a Belgian Roman Catholic priest and noted astronomer, went further in 1931, and suggested that if the present position and velocity of these known galaxies were projected backward in time, these objects, all the mass in the universe, and even space itself would come down to a single point. The bold conclusion is that the entire universe started at a precise date and time at a dimensionless point in the far but calculable past.

This theory of the beginning of everything was named the big bang during a radio broadcast in 1949 when the distinguished British astronomer Fred Hoyle (1915–2001) blurted it out. He was disparaging this theory of creation while advancing his own steady state theory. Since then, the big bang theory has been strengthened and confirmed with more observations, most notably the discovery of the *cosmic microwave background radiation* in 1964.

This cosmic radiation was discovered by Robert W. Wilson (1936–   ) and Arno Penzias (1933–   ) while employed at the Bell Labs in Holmdel,

Robert W. Wilson (right) and Arno Penzias in front of their microwave radio antenna at Bell Labs, Holmdel, New Jersey, on October 18, 1978 *(AP Images)*

New Jersey. Working with a new type of microwave antenna, they found a curious noise coming over their radio receiver at 160.2 gigahertz. This noise was coming from all directions, so it could not be coming from anything near. With effort, they found that it was coming from outer space. In addition, it was coming from everywhere in outer space. They had found evidence of an event that had been predicted to happen 379,000 years after the moment of creation, when matter and radiation decoupled and microwave radiation burst forth. The event has been echoing through the universe for billions of years since, bouncing off solid surfaces, and the pulse has become faint, but it still exists. On a microwave receiver, it is a hiss. It sounds like static.

Wilson and Penzias were awarded the Nobel Prize in physics in 1978 for their discovery, and many subsequent experiments have bolstered this finding and the concept of a big bang creation.

All indications are that the universe began with an enormous explosion and hence the name big bang. In the very beginning, the universe was crowded into a single point, or singularity, having infinite density and temperature. Such a situation was bound to expand, quite rapidly, in a cosmic inflation. As the universe expanded, it cooled. The most fundamental particles of matter, quarks and gluons, formed an extremely high-temperature plasma. Subatomic particles, such as electrons and anti-electrons, were being created and then destroyed as quarks and gluons collided. After one-millionth of a second, the temperature of the universe had cooled down by expansion sufficiently to allow protons and neutrons to be assembled and stay together. A mass annihilation followed, as particles and their antiparticle equivalents found each other and mutually destructed. Fortunately, there were slightly fewer antiparticles produced than regular particles. Although only one out of every 10 billion particles survived, there was enough material left to build an entire universe. At that time, most of the cosmos consisted of gamma rays, with some remaining protons, neutrons, and electrons.

After a few minutes into creation, space was still growing larger, and the temperature had dropped precipitously, to about a billion degrees. The density of matter was about that of air at sea level. At this lowered temperature, the protons could be considered hydrogen in the plasma state, with electrons simply part of the soup and not constrained to orbits. The conditions were sufficiently hot and dense to promote fusion, and hydrogen fused with hydrogen and with free neutrons to form deuterium, or hydrogen-2, and helium. A deuterium nucleus consists of a proton plus an attached neutron, and helium-4 is two protons plus two neutrons. Thus began big bang nucleosynthesis, in which heavier elements were made from the lightest element, hydrogen, under what seemed to be low-temperature conditions at the time.

Still, the universe expanded, as it still expands. After 379,000 years, the universe had expanded to a size that allowed its temperature to drop to a non-plasma temperature, or about 4,940°F (3,000°K). At this point, electrons started to settle into orbits around atomic nuclei. The energy transition as electrons snapped into quantum orbits caused an enormous, universe-sized pulse of microwaves. It is that pulse that can still be heard. The universe at that point consisted of matter as we know it, and the temperature continued to drop.

STELLAR NUCLEOSYNTHESIS: MAKING ELEMENTS BY NUCLEAR FUSION

Although there was excellent theoretical work indicating that elements heavier than helium were being made in stars, there was no direct proof of these theories until the early 1950s, when the characteristic spectrum of the element technetium was discovered in the visible light from a red giant star.

Technetium is the only naturally occurring element of which there is not a trace on Earth. The reason is that all variations or nuclides of technetium are radioactive, and the longest-lived example, technetium-97, has a half-life of 2.6 million years. Earth is about 4.5 billion years old, so any technetium that was here at the beginning has long since decayed away. There are 17 nuclides of technetium, and the shortest-lived example has a half-life of 0.83 seconds. Given the longest possible technetium half-life of 2.6 million years, seeing technetium in starlight means that the technetium had to have been constructed on that star. The average red giant star is at least 10 billion years old, so any technetium residue in the original star would have disappeared by radioactive decay. There were many heavy elements discovered in starlight spectra beforehand, but the finding of technetium was indisputable evidence of nucleosynthesis in the fusion furnace of a star.

Working at Cornell University beginning in 1935, the theoretical physicist Hans Bethe had carefully developed a scheme by which helium-4 can be produced from hydrogen-1 in a large star. In this process, carbon, nitrogen, and oxygen are used as catalysts (called the *CNO cycle*.) A star the size of the Sun uses the proton-proton fusion reaction for most helium production, with less than 2 percent of this conversion being due to the CNO cycle. In larger stars, from those that are 1.3 times the size of the Sun to those that are hundreds of times bigger, the CNO cycle becomes a dominant process.

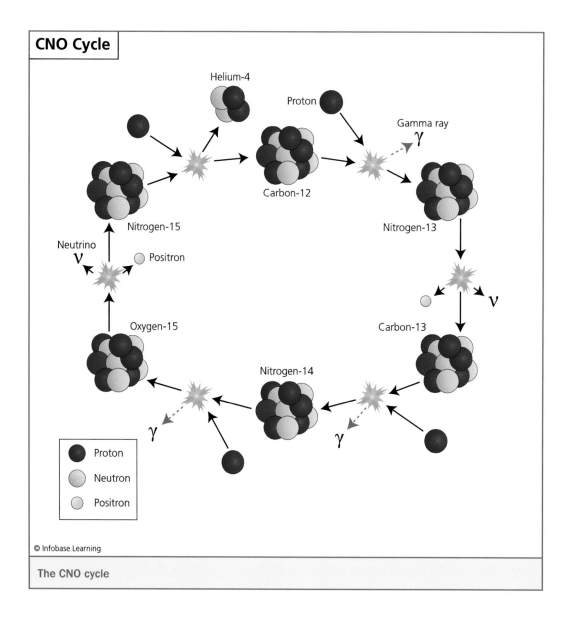

CNO Cycle

Helium-4

Proton

Gamma ray
γ

Carbon-12

Nitrogen-15

Nitrogen-13

Neutrino
ν

Positron

Oxygen-15

ν

Carbon-13

Nitrogen-14

γ

γ

● Proton

○ Neutron

○ Positron

© Infobase Learning

The CNO cycle

HANS BETHE (1906–2005):
THE SUPREME PROBLEM SOLVER

Hans Bethe, the theoretical physicist who unlocked the complexities of stellar nucleosynthesis, was born in Strasbourg, Germany, to a Jewish mother and a Christian father. He was raised a Christian and showed early aptitude in science. He went on to study physics at the Johann Wolfgang Goethe University in Frankfurt and earned his doctorate in physics at the University of Munich. He did postdoctorate work at Cambridge, England, and with Enrico Fermi's nuclear research group in Rome, Italy.

He was hired as an instructor at the University of Tübingen in Germany, but in 1933 the German government implemented official measures against Jews in professional positions. Although he was technically a Lutheran, he was fired because of his mother's background. Seeing that things would only get worse, he moved to England and never looked back at his homeland.

He stayed for a year at the University of Bristol, working on a comprehensive theory explaining the properties of deuterium, one of two heavy isotopes of hydrogen. In 1935, he was offered a faculty position at Cornell University in the United States, where the elite nuclear physicists from Europe seemed to be congregating in the years before World War II. He flourished there, becoming one of the leading theoretical physicists of his generation and helping to put Cornell on the map. There, from 1935 until 1938, he studied something that had never been seen in a laboratory, *nuclear fusion* reactions. He formulated the CNO cycle with Carl Friedrich von Weizsäcker (1912–2007), who had remained in Germany, and they published together.

In 1942, Bethe and every other physicist or almost physicist in the United States was recruited for work on the Manhattan Project to quickly develop nuclear weapons under wartime conditions. He was immediately named director of the theoretical division at the Los Alamos weapons design laboratory. At Los Alamos during the war, he spearheaded some critical work, including calculating the critical mass of uranium-235 and the ultimate explosive yield of the bomb designs. He gained a strong reputation as "the Supreme Problem Solver."

The bomb development project was entirely successful and, although he was initially opposed to the development of an even larger, hydrogen fusion bomb after

(continues)

(continued) _____

the war, he signed up for the project and saw it through. He was philosophically torn, being a pacifist at his core, hoping for world disarmament, and yet seeing a need for the United States to have a fusion weapon before other powers were able to develop this type of superweapon. At the age of 88, he wrote an open letter urging all scientists to cease work on any aspect of nuclear weapons development or manufacture.

Always busy, he continued vigorous research on supernovae, black holes, and theoretical astrophysics into his late 90s. In 1967, Bethe was awarded the Nobel Prize in physics "for his contributions to the theory of nuclear reactions, especially his discoveries concerning the energy production in stars." It was the first Nobel Prize awarded for work in fusion research.

In the CNO cycle (see figure on page 18), a hydrogen-1 nucleus fuses with a previously built carbon-12 nucleus. The resulting energy occurs as a gamma ray. The resulting nitrogen-13 nucleus then undergoes radioactive decay, emitting a positron and a neutrino. The result is a carbon-13 nucleus, which fuses with another hydrogen-1 nucleus, or proton. The resulting energy is expelled as a gamma ray, and the product of the reaction is a nitrogen-14 nucleus. The nitrogen-14 fuses with yet another hydrogen-1, forming oxygen-15 and another energetic gamma ray. The oxygen-15 nucleus then beta decays into nitrogen-15, releasing a positron and a neutrino. A hydrogen-1 then fuses with the nitrogen-15 to produce two offspring, an energetic helium-4 and a carbon-12, bringing this busy process full circle. The two positrons, or antielectrons, produced in this ring of actions find two electrons to destroy, and pure energy is released. The released neutrinos escape the star in straight, undisturbed paths and are considered unrecoverable energy.

This is the most straightforward of three known CNO cycles, and there are many more, increasingly complex ways that suns light the night sky by fusion. This means of converting the mass of the universe gradually into pure energy also happens to assemble the building blocks, the elements of which all planets, asteroids, and comets are composed.

EXPLOSIVE NUCLEOSYNTHESIS: MAKING THE HEAVIEST ELEMENTS

On February 24, 1987, two astronomers at the Las Campañas Observatory in Chile were looking at the night sky at a nearby dwarf galaxy named the Large Magellanic Cloud. It is only 168,000 light-years away. They saw an unusual burst of light, visible to the naked eye. The discovery was confirmed by an observer in New Zealand, and it was assigned a name, *SN1987A,* or the first supernova discovered in 1987. It was the first observation of an up-close supernova in the history of modern astronomy, and it gave positive confirmation of a long-held theory, that heavy elements are made only when a star explodes at the end of its life.

Unlike a star the size of the Sun, an old, large red giant star can make elements heavier than iron by neutron capture and beta decay, up to an element as heavy as lead, but only in an inefficient process resulting in only trace quantities. If the universe relied on a normal star lifetime to build all the elements, then there would be no uranium or thorium, and nuclear fission power would not be possible.

A nova is a rare phenomenon, as most stars simply dim out and collapse at the end of life. The materials for fusion are used up after having kept the star hot and glowing for more than 10 billion years. But occasionally the conditions for a nova exist. If a white dwarf star of sufficient mass has burned up all the fusing material at its center, then it can collapse on itself as the outward pressure from the fusion stops. This increases the core temperature significantly, and carbon atoms in the star begin to fuse together. Most of the matter in the star is ignited into fusion by the energetic carbon fusion, and the entire star becomes unbound in a nova explosion. A shock wave propagates through the gas surrounding the star, traveling at about 3 percent of the speed of light. The destruction of the entire star takes only a few seconds. In an even rarer event, a supernova, the core of a collapsing star fuses into iron in an explosion of such intensity, the light emitted is greater than the combined light from all other billions of stars in its home galaxy.

Supernova SN1987A actually happened 168,000 years ago, but it took the light and radiation that long to reach the observatory in Chile. Gamma rays were detected coming from this event, identified by energy as being the characteristic radiation from the decay of cobalt-56 and cobalt-57. Cobalt decays into iron, and these isotopes of cobalt have half-lives of only about a year. The sudden appearance of the cobalt specimens confirmed

A composite image of the Crab Nebula, the remains of a supernova explosion observed by Chinese and Arab observers in 1054. The bright white dot at the center is a neutron star that remained after the event. (*NASA Jet Propulsion Laboratory*)

that heavy elements are made in supernovae. Further study of the stardust grains left over in the surrounding space where once a star had been confirmed this finding, particularly with the discovery of titanium-44 in the debris.

A supernova occurs in a star that has shrunken to the size of the Earth, with a material density of about 36,000 pounds per cubic inch (1 MT per cc). While this process produces a great deal of energy in a short time, these conditions would prove difficult to attain on a continuous basis in an Earth-bound power plant. Still, the quest for fusion power has a long and colorful history.

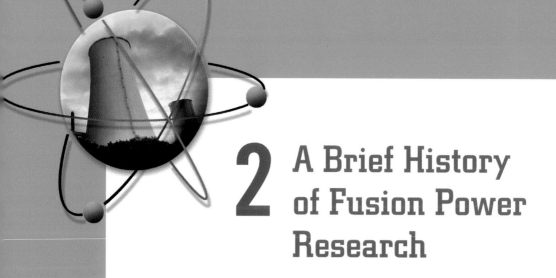

2 A Brief History of Fusion Power Research

As is the case with fission power, the public was made aware of nuclear fusion by the explosion of a very large bomb. In 1952, the first hydrogen bomb, or thermonuclear device, was detonated experimentally by the United States, and it was difficult to hide this secret from the population of the world. A Hollywood motion picture was made of the entire test, its preparation, execution, and aftermath, and it was shown on television for several weeks. With a jolt, everyone near a news source learned of a new energy source, even more powerful than fission. The island on which the test was conducted vanished without a trace during the explosion.

The promised properties of this new energy source were even better than the advantages of fission power by uranium. There were, however, details to be worked out. Work continues on to this day.

EARLY INDICATIONS AND THEORIES OF NUCLEAR FUSION

The first theoretical indication that two atomic nuclei could fuse was formulated by the unlikely matchup of Fritz Houtermans (1903–66), a German physicist and avowed communist, and Robert Atkinson (1898–1982), a British astronomer and dedicated clock-maker. The year of their collaboration was 1929. Houtermans was at the University of Göttingen in Germany, a hotbed of theoretical nuclear physics, and Atkinson happened to be on a Rockefeller Travelling Scholarship, working on his Ph.D. at the same university. Together they measured the masses of low-mass elements, beginning with hydrogen and helium. An atom of helium-4 is composed of two protons and two neutrons, but it did not weigh quite as much as two hydrogen-2, or deuterium, atoms, which are each one proton and one neutron. Using Einstein's discovery of the mass-energy equivalence, they predicted that a great deal of energy would be released if two deuterium atoms could be fused into one helium-4. The slight deficit in mass found in the helium-4 would turn directly into energy, probably under the conditions in the center of a star.

The first experimental confirmation of this fine theory was performed by Mark Oliphant (1901–2000), an Australian physicist at the Cavendish Laboratory, University of Cambridge, England, in 1932. He discovered the light isotopes helium-3 and *tritium,* or hydrogen-3. He was also very interested in particle accelerators, used to crash together two ionized atoms or atoms and charged particles in a vacuum using high voltage potentials. His natural inclination was to crash ionized examples of his newly discovered atoms into each other at high speed.

In doing his tests, Oliphant managed to fuse together nuclei, two at a time making one, heavier nucleus. Deuterium, for example, would fuse with tritium, making helium-4 plus one free, energetic neutron. Terrestrial, small-scale fusion giving a net energy release was therefore established. The temperature and pressure at the center of a star were not necessary. However, running up the high voltage on the accelerator required much more energy than was recoverable from the reaction. This negative energy production would prove to be a stumbling block to harnessing controlled fusion for the foreseeable future.

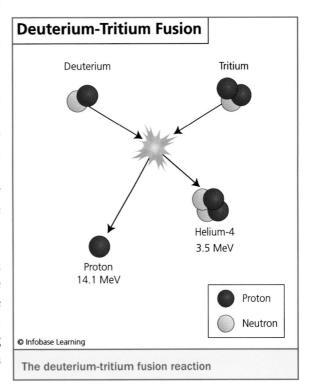

The deuterium-tritium fusion reaction

BUILDING A FUSION REACTOR IN 1938

Pioneering work in fusion energy can come from unexpected places. The first attempt to fuse hydrogen plasma in a toroidal magnetic field was performed in 1938 at the Langley Memorial Aeronautical Laboratory in Hampton, Virginia. It was the flagship facility of the National Advisory Committee for Aeronautics, or NACA, the predecessor of NASA, and home of a large, variable-density wind tunnel.

Arthur "Arky" Kantowitz (1913–2008), a young physicist recently graduated from Columbia University, had been hired to assist in improving the performance of aircraft wings. His boss was Eastman Jacobs (1902–87), a free-spirited mechanical engineer with a bachelor's degree from the University of California, Berkeley. Krantowitz had just read in a magazine article that Westinghouse had bought a Van de Graaff high-voltage generator, and he felt sure that they were going to use it to duplicate the exciting nuclear fusion experiments reported from the Cavendish Laboratory in England. It was good to duplicate an experiment, but it was even better to improve an experiment for enhanced results, and Arky Krantowitz was brimming with ideas. He had no trouble fascinating his section head, Jacobs, with the concept of building a fusion device.

The initial problem was that fusion research was nowhere near the mission of NACA and the wind tunnel facility, so they had to come up with an obscuring name for the project. They called it a Diffusion Inhibitor and convinced the management to allow an expenditure of $5,000.

The design of the experimental device was ingenious, and nothing like it would be built again until the early 1950s. The researchers wished to duplicate fusion conditions at the center of the Sun, and the way to do that was to forcibly ionize some

In 1939, Hans Bethe published his CNO cycle paper, and this significantly advanced the scientific understanding of nuclear fusion.

A SUCCESSFUL APPLICATION OF FUSION THEORY: THE HYDROGEN BOMB

Immediately after the close of World War II, all available research funding in advanced energy systems went into controlled fission research, with eventual emphasis on submarine propulsion. Fusion was too far behind at this early juncture, and fission was proven.

In 1949, the Soviet Union successfully tested a fission-powered nuclear weapon, and concern over this loss of military advantage caused the start hydrogen gas into plasma and contain it in a magnetic field. To ensure that there was not a leak-point for the plasma, they made the magnetic field in a donut shape, or a torus. Ionization energy was pumped into the hydrogen gas using a 150-watt radio transmitter. They understood that a deuterium-deuterium fusion would be advantageous, but unfortunately in 1938 there was no ready supply of deuterium. They would have to make do with ordinary hydrogen gas. They hoped to heat the hydrogen plasma to at least 10 million degrees and detect some X-ray emissions as a beginning experiment. The fusion reactor was made of half-inch aircraft aluminum plates, welded together and wound on the outside with water-cooled copper wire.

The experiment had to be conducted in secrecy, not for national security, but so that the upper management would not find out what they were doing. The equipment could only be turned on after midnight, so that the current-draw would not dim the lights in the neighborhood and give them away.

Success was not immediate, and the problems of plasma containment that would haunt every other fusion experiment in the next 70 years visited early. Krantowitz tried holding the building circuit breaker closed so they could apply more power to the plasma, but evidence of radiation did not show up on photographic film in the chamber, although a characteristically blue plasma glow could be seen through a view port on the torus.

Just as the two researchers were beginning to learn the problems with plasma, the research director happened by and tripped over their unorthodox experiment. He listened patiently to an explanation and then cancelled the program on the spot. A bold first step in fusion power research thus came to a premature end. Krantowitz drifted away to found the Avco-Everett Research Lab in Everett, Massachusetts, and Eastman opened a restaurant in Malibu, California.

of another crash program to develop an even bigger bomb. Under cold war footing, the United States embarked on a large, expensive effort to develop fusion-powered weapons. Using the process that powers the universe, such devices were theoretically capable of delivering 1,000 times more power than the bombs that had completely destroyed two Japanese cities at the end of the war. For this work, the fusion phenomenon would have to be studied more closely than ever before, and the experiments would be expensive.

Edward Teller (1908–2003), a brilliant theoretical physicist from Hungary, had been an early proponent of such research. With the new push for the hydrogen bomb, he was able to go back to work at his wartime job, designing bombs at the weapons lab in Los Alamos, New Mexico. As the complexity and the sweep of the work increased, he was able to found a new, additional weapons lab at Livermore, California.

There was nothing simple about expanding the fission-based atomic bomb into a larger fusion bomb. Early designs used a fission bomb as a detonator, colocating a deuterium-tritium gas mixture with the fissioning plutonium-239, in layers. This design, code-named "Alarm Clock," looked good on paper, but not under digital simulation. A new type of analysis hardware was available after the war, having been developed in secrecy. It was the digital computer, and the first use of this technology was to simulate Teller's bomb designs mathematically. The "Alarm Clock" failed to detonate in the computer simulations. The gas mixture seemed to cool off too quickly to achieve explosive fusion.

Just when the hydrogen bomb design began to look hopeless, Stanislaw Ulam (1909–84), a mathematician from Poland, stepped in to rescue the concept. His design was to separate the fission bomb from the fusion module, instead of having them on top of each other. This would give the fission bomb time to develop a complete explosion. The shock wave from this violent event would then be sufficient to collapse a nearby volume of deuterium-tritium, causing it to be compressed and heated to the point of fusion.

Teller was intrigued by the idea, but he took it a step further. The first thing out of the atomic bomb, in the first microseconds after its ignition started, would not be a shock wave or even gamma rays or neutrons. The first casualties in an atomic bomb explosion are the electron orbits of the fissioning atoms. When the electrons are boiled off, powerful X-rays are generated. This extreme, pulsed wave of X-ray energy would be the force that would compress the neighboring deuterium-tritium and cause it to fuse. Computer simulations confirmed that this bomb would explode with more than adequate force.

In 1952, a military task force was gathered in the middle of the Pacific Ocean at Eniwetok Atoll to conduct an enormous experiment. A 300-foot (91-m) radio and television tower was erected on Eugelab Island and, under it, was a cube-shaped building. In the building was assembled the *Teller-Ulam device,* consisting of an atomic bomb atop a two-story Dewar flask of a liquefied mixture of deuterium and tritium gas. Next to it was a cryo-refrigeration plant, keeping the hydrogen isotopes at liquid temperatures of less then –473°F (–253°C). The bomb detonation was set up as a fully instrumented scientific experiment, with a helium-filled tunnel, 3.5 miles (5.6 km) long, conducting light from the explosion to high-speed

Hydrogen Bomb

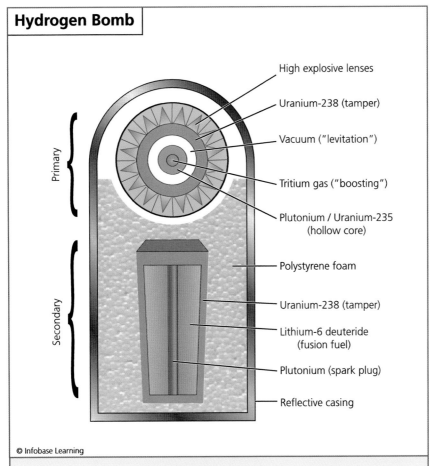

High explosive lenses

Uranium-238 (tamper)

Vacuum ("levitation")

Tritium gas ("boosting")

Plutonium / Uranium-235 (hollow core)

Polystyrene foam

Uranium-238 (tamper)

Lithium-6 deuteride (fusion fuel)

Plutonium (spark plug)

Reflective casing

Primary

Secondary

© Infobase Learning

The Teller-Ulam hydrogen bomb design. The primary, a conventional atomic bomb, goes off first, and then X-ray emissions from the primary reflect off the casing and compress the secondary with extreme pressure. Excess neutrons released in the A-bomb explosion activate the lithium-6 into tritium, and it fuses with the deuterium. As a final kick, the plutonium rod in the middle of the secondary fissions explosively.

motion picture cameras on a nearby island. The official code word for the test was "Ivy Mike."

On November 1, 1952, at 7:15 A.M. local time, the device was detonated, and performance exceeded all expectations. It made a fireball 3.5 miles (5.2 km) wide and a mushroom cloud 57,000 feet (17 km) tall. It proved that fusion producing significant energy, at a level of about 12 megatons of TNT (5×10^{16} j), could be achieved on Earth, without the mass of a star pressing down on it. An improved hydrogen bomb, ROMEO, using a dry powder

A hydrogen bomb explosion, Operation Castle, ROMEO event, detonated from a barge near Bikini Atoll in the Pacific Ocean on March 26, 1954 *(National Nuclear Security Administration/Nevada Site Office)*

of lithium deuteride as the active ingredient, was exploded successfully on March 26, 1954. These events lasted less than a second, and a more controlled, sustained mode of fusion would be necessary for power production.

THE SPARK IN ARGENTINA THAT STARTED FUSION RESEARCH

With the United States and the Soviet Union now fully absorbed in a race to achieve a fusion-powered explosive, insignificant attention was paid to a quest for a system that could generate power continuously with fusion. Out of the blue, on March 24, 1951, the government of Argentina announced that on February 16, "thermonuclear reactions under controlled conditions were performed on a technical scale." A scientist in Argentina had managed to produce significant energy by fusion in a device the size of a milk bottle. The announcement shocked the scientific communities in the United States and Europe. They seemed to have been caught off guard.

After World War II, Argentina was in excellent economic condition, with a large trade surplus and an ambitious general named Juan Perón (1895–1974) as president. With the intent to modernize Argentina with scientific and industrial innovation, he hired a German scientist Ronald Richter (1909–91) in 1948 to pursue an intriguing idea. Richter claimed to have worked on nuclear fusion in Germany and said he could induce fusion in deuterium by the use of shock waves from high-velocity particles. His process could produce controlled nuclear fusion in enormous quantities and, of greatest importance, it was something that the United States and the entire continent of Europe could not accomplish.

Ruins of the main laboratory on Isla Huemul, Argentina *(Guy Walters)*

Richter was given a laboratory site on the secluded Isla Huemul in the middle of Nahuel Huapi Lake in the mountains of western Argentina and a blank check for a budget. Seven million dollars, a lot of money at the time, was poured into the project. A building complex was constructed on the island with a thick-walled concrete laboratory at the center. Work progressed in utter secrecy, with no published theories or experimental findings and certainly no peer review. Graduate physics students were not hired to work on the project, as had been planned, and Richter seemed content to work alone. By early 1951, Richter reported positive results to his sponsor, Juan Perón, and an announcement was made in triumph.

Unfortunately, Richter had not actually achieved any fusion in his apparatus, as eventually became obvious to Argentine physicists finally allowed to see his setup. Richter's work was a shambles, with no coherent record of what exactly he had been doing. His supposedly successful

An interior view of the main laboratory building in the Huemul ruins *(Guy Walters)*

experiments could not be replicated, and even cursory examination of his original theories revealed some obvious gaps and unrealizable assumptions. Richter was thrown in jail for seven days for contempt of congress, the laboratory on the island was eventually abandoned, and his *Huemul Project* slid quietly into obscurity. Juan Perón lost his leadership position in a military coup on September 19, 1955.

This embarrassing incident, however, had a stirring effect on scientists worldwide. It had uncovered a need in the human psyche for the promise of a perfectly clean, nonpolluting power source that had no end of available fuel. With all the emphasis on the development of fusion weapons, science had ignored this even more important aspect of fusion. Interest peaked in the United States, Britain, and the Soviet Union for controlled fusion, and it became a technical race.

MAGNETIC PINCH DEVICES SHOW PROMISE AND DISAPPOINTMENT

Inspired by the failed Argentine fusion project, an American physicist Lyman Spitzer (1914–97) proposed a bold plan. Spitzer was an astronomy professor at Princeton University in New Jersey. His specialty was the study of superhot interstellar gases and plasmas, such as are found in stars. His immediate thought was that controlled, sustained fusion would occur in hydrogen plasma, and he had ideas for how to produce power using it. He wrote a proposal to the newly formed Atomic Energy Commission (AEC) to build three machines to study plasma, at a cost of about $1 million. He received funding from the AEC, assembled a small group of experts, and commandeered a small building on the outskirts of the Princeton campus.

The building, once used to house rabbits for animal experiments, was called the "Rabbit Hutch" and the project was named "Matterhorn." As was the case of any government-funded project that had the word *fusion* in its descriptor, even though it did not involve weapons research Matterhorn was classified SECRET. No information could be leaked to the public, and the windows on the Rabbit Hutch had to be painted over.

Spitzer's vision was to build a tube to contain a small amount of hydrogen gas. The gas would be "pinched" to high density and temperature in a plasma state by an intense magnetic field. The magnetic field would be created by wire wrapped around the tube with electricity flowing through it. The plan was elegantly simple, but there was a problem. As it was being

Lyman Spitzer and his Model A stellarator *(Princeton Plasma Physics Laboratory)*

pinched, plasma would shoot out both ends of the device, like toothpaste out of a stepped-on tube. Spitzer had an elegant solution to this problem. The tube, which was 12 feet (3.7 m) long, was bent into a figure eight. The tube therefore had no ends, but was one continuous plasma path.

Spitzer named it the *stellarator,* and quick results were expected. A plasma temperature of 5 million degrees was theoretically possible. By fall 1952, the Model A stellarator was ready for tests. The team extinguished the lights in the building and turned the device on. Spitzer could see a faint, purple glow in the observation ports, but not for long. The machine did make plasma, but it lasted only about a thousandth of a second. The plasma reached about half a million degrees before it escaped the magnetic field and hit the inner wall of the tube. As soon as it hit the wall, the plasma extinguished.

This was a setback, but Spitzer felt certain that the problem was the small size of the stellarator. By 1954, the team had assembled the Model

B stellarator, larger, more powerful, and with many improvements. It achieved 1 million degrees inside, but still only for a thousandth of a second. The problems in the nascent science of plasma physics were deeper than had been realized. Plasma behaved wildly, surging and twisting to release itself from the constraints of the magnetic field. A larger machine was needed, and a Model C stellarator was constructed. This time it had $20 million in funding, a new building, and its own power station. Still, successful generation of a sustained plasma seemed elusive, and no fusion reaction was detected. In 1961, the SECRET classification was completely dropped, Spitzer resigned, and the project was renamed the *Princeton Plasma Physics Laboratory*. Several different configurations of stellarator were tried by a new generation of optimistic scientists, including the torsatron, the heliotron, the heliac, and the helias. Stellarator projects were conducted in Germany and Japan. None was able to achieve significant fusion.

The last stellarator project to be funded at Princeton was the *National Compact Stellarator Experiment (NCSX)*, but the work was cancelled in 2008 due to a failure to meet budgetary constraints. At this point, the stellarator concept has yet to achieve a stable, fusing plasma.

British scientists were also inspired by the Argentine fusion debacle to pursue a plasma pinch. Their funding was not immediate, but in 1954 the British government decided to fund a push for a fusion power reactor. Money went to the *Atomic Energy Research Establishment (AERE)* at Harwell, Oxfordshire. This main research center for atomic energy was headed by Sir John Cockcroft (1897–1967), winner of the Nobel Prize in physics in 1951 for his work with particle accelerators.

The British concept of controlled fusion also involved plasma, but their method of pinching it was unique. Instead of applying an external magnetic field, a pinching field would be generated inside the plasma itself. Plasma conducts electricity. It is, in fact, a superconducting material, imposing no resistance on an electrical current. The British plan was to shoot a high-voltage current through an established plasma stream, causing it to generate a strong magnetic field. The hydrogen plasma would attract itself, collapsing into a hot, dense knot at the center, fusing into helium and releasing energy.

The first attempt was a proof-of-concept experiment and not a full power plant. It was named the *Zero-Energy Thermonuclear Assembly (ZETA)*. The device was designed to run only for a fraction of a second, but it was the largest fusion device in the world, weighing 120 tons (132 MT). The power

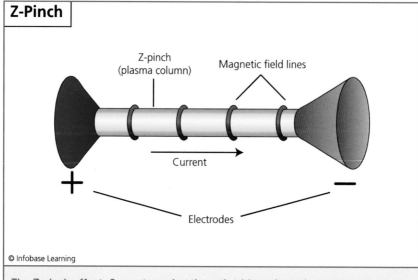

Z-Pinch

The Z-pinch effect. Current running through tritium-deuterium gas produces a compressing magnetic field. Unfortunately, the reaction spills out both ends of the tube.

for the pinch had to be stored in a room-sized bank of capacitors. When fully charged with electricity, energy in the capacitors was switched into the ZETA. There was a momentary blue flash, and the foundation of the building shook as the magnetic force peaked. In these experiments, the scientists found that the plasma lost initial stability quickly, buckled, twisted, and broke up, quenching any reaction. Applying greater power simply made it break up faster. However, after four years of tweaking and adjusting the plasma chamber, neutrons were detected coming from inside the machine, and this was interpreted as evidence of nuclear fusion. Deuterium nuclei were apparently knocking together in pairs to produce single nuclei of helium-3, with the excess neutrons from the reaction escaping the plasma chamber. A scale-up of this apparatus could produce usable energy.

Feeling confident of the results and genuinely excited, Cockcroft called a press conference on January 24, 1958, to announce the world's first successful achievement of plasma fusion in a laboratory. He announced that the hitherto secret ZETA device had produced plasma temperature of 5 million degrees for up to three-thousandths of a second, noting that this was far better than anything the Americans had produced. He further announced that fusion reactions were taking place in the heated plasma.

The news media of Great Britain exploded with the tidings. Headlines the next morning were "ZETA SPELLS H-POWER EVERLASTING, BRITAIN'S H-MEN MAKE A SUN," and "LIMITLESS FUEL FOR MILLIONS OF YEARS." The *Daily Telegraph* ran the headline "U.S. ADMITS THAT BRITAIN HAS THE LEAD."

The euphoria lasted about four months, at which time it was discovered that the neutrons produced by ZETA were not coming from nuclear fusion. The neutrons were being released when the plasma hit the wall, as it had in the stellarator only at a lower energy. No energy was being produced. The world's science community took no delight in the crash of the British exuberance. At Harwell, ZETA was used for experiments for another 10 years, exploring problems with plasma

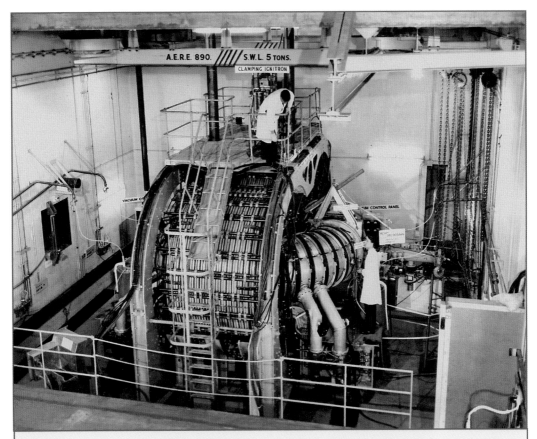

Zeta reactor with vacuum spectrograph fitted, ca. 1956 (© SSPL/Science Museum/The Image Works)

containment. No further press announcements of a fusion break-through were made.

Soviet scientists were also impressed and inspired by the Huemul Project announcement from Argentina. Although they were striving to design a thermonuclear weapon at the time, they became excited at the concept of building a fusion reactor for power production. Igor "the Beard" Kurchatov (1903–60) was head of the Soviet atomic bomb project. Kurchatov had formed the first Soviet nuclear science research team at the Physico-Technical Institute in Petrograd, Russia, in 1932 and had built the Soviet Union's first cyclotron in Leningrad in 1937. He took over the

Tokamak I

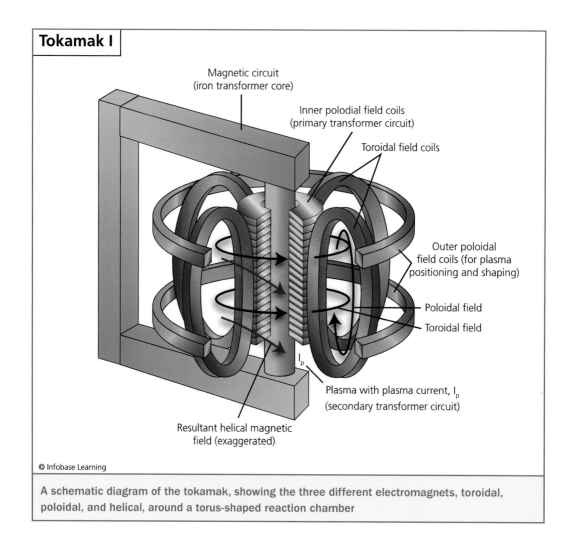

A schematic diagram of the tokamak, showing the three different electromagnets, toroidal, poloidal, and helical, around a torus-shaped reaction chamber

town of Sorov, Russia, renamed it Arzamas-16, populated it with the best scientists in the Soviet Union, and successfully built and tested a fission-based nuclear weapon in 1949. This accomplishment brought the Soviets to equal weapons status with the United States, but both powers were then saddled with coming up with a more powerful, fusion-driven device. In 1951, Kurchatov took a slight detour, directing his best theoreticians, Andrei Sakharov (1921–89) and Igor Tamm (1895–1971), to come up with something to beat the British and the Americans in a race for controlled fusion power.

The two scientists began with an assumption that hydrogen would have to fuse in the hot plasma state and therefore a plasma must be contained in a small space and kept away from any mechanical interference, such as the walls of the container. The logical thing to do was to establish a strong magnetic field in the container. Furthermore, the plasma could be given no avenue of escape, and this meant there could be no end points to the magnetic field. The same configuration occurred to all who started from scratch on this problem. The machine would have to be donut shaped, or a torus, wrapping the contained plasma around on itself.

This same concept was being pursued by both the other teams, but the Americans were using an external set of coils to establish a magnetic field. The British team was using an electrical current through the plasma to provide magnetism. The Soviets had a more advanced idea. They would use both methods in the same device.

The design was first named the magnetic thermonuclear reactor, but it would soon take its most famous name, the *tokamak*. The word was a Russian acronym, from *toroidal'naya **kamera** s **magnitnymi katushkami**,* or toroidal chamber with magnetic coils. To the surprise of the researchers, plasmas in the device proved unstable and impossible to control. Although the tokamak seemed superior to anything else for plasma control and temperatures, it was still going to be a long effort to achieve true, controlled fusion. Starting with an experimental device named T-1, the torus was not very large. The external radius was only 31.5 inches (0.8 m), and the internal radius, or the donut hole, was 5.1 inches (0.13 m). Initial indications were that a larger device would be needed.

Kurchatov's researchers moved through successive improvements, to the T-2, the T-3, and to the largest version yet, the T-4. T-4 occupied the entire floor of a building at the Kurchatov Institute. It was something of a breakthrough, as it produced the first quasi-stationary thermonuclear fusion reaction. In early 1968, after more than a decade of tokamak experimentation, plasma in the T-4 reached temperatures of up to twenty-million degrees for as long as 20 thousandths of a second, and this outperformed anything else that had been tried. Results were announced at the third International Atomic Energy Agency International Conference on Plasma Physics and Controlled Nuclear Fusion, held in Novosibirsk, Russia, in July of the same year. It was a rightly earned triumph for the Soviet Union in plasma and controlled fusion research.

Since then, the tokamak has become the most studied and copied technology in the history of fusion research. A total of 208 tokamaks have been built and tested around the world, in Europe, Asia, Australia, and North America. The Russians have been operating the T-10 tokamak since 1975. Many plasma containment and fusion power records have been broken with tokamaks. In 1993, scientists using the TFTR tokamak at Prince-

The Tokamak Thermonuclear Facility at the Kurchatov Atomic Energy Institute, taken on March 2, 1989 (© RIA Novosti/Alamy)

ton University, experimenting with a 50–50 mix of deuterium and tritium gases, produced as much as 10 megawatts of power by fusion, momentarily. In 1996, a record duration of two minutes of operation was achieved at the *Tore Supra* tokamak in Bouches-du-Rhône, France, using *superconducting magnets*. In 1997, the *Joint European Torus (JET)* tokamak in Culham, Oxfordshire, England, produced 16 megawatts of fusion power. This is a current world record, achieved using an explosively squeezed plasma technique. It was still necessary to provide considerably more than 16 megawatts of power to initiate the fusion event.

Although a break-even, usable power production has yet to be reported, the tokamak remains a promising fusion reactor design, and reactors of this basic design are still being built.

LASER FUSION TAKES THE LEAD

A major problem with the magnetic containment of fusion plasma seems to be the duration of a stable plasma. It is an accomplishment to generate power with a plasma machine, but until that power can be maintained for months or years, this does not seem a practical energy source. There is another direction from which to consider making usable fusion power. If it is accepted that power can only be made in millisecond bursts, then design a reactor that operates with a continuous stream of contained explosions, one after another. In the 1960s, 10 years after development began on magnetic fusion, laser fusion, igniting tiny thermonuclear explosions in deuterium-tritium ice, became an even larger subject of experimentation. This research track is more properly known as *inertial confinement fusion (ICF)*. Billions of dollars of research funding have gone into this branch of fusion reactor technology.

Thought in this direction began shortly after the invention of the laser, or light amplification by stimulated emission of radiation. A laser is a device that produces an intense, narrow beam of light in one frequency or color. Laser light can be so intense that it actually acts as a force, imposing pressure and heat. These are two qualities that can be used to induce hydrogen isotopes to fuse, as they do in the center of the Sun. If a small, solidified pellet of deuterium-tritium could be brought to simultaneously high temperature and high density, then fusion would occur in the center of the pellet. Laser light hitting it from all directions could do this, but the pellet would quickly explode and lose its ability to fuse. To generate

GENERAL ELECTRIC'S "FUSION ON EARTH" EXHIBIT AT THE NEW YORK WORLD'S FAIR

In 1964 and 1965, a grand spectacle of a world's fair was held in Flushing Meadows–Corona Park, in the borough of Queens, New York. Hailed as a universal and international exposition, the fair's theme was "Peace Through Understanding," dedicated to "Man's Achievement on a Shrinking Globe in an Expanding Universe."

Although some European countries, Australia, Canada, and the Soviet Union boycotted the fair on official sanctioning issues, many smaller nations set up pavilions and participated enthusiastically. The fair was, however, thick with American culture and technology. The age of space exploration, with its race for the first Moon landing, was an unspoken but ever-present motif, and the fair was a showcase for American industry. Many innovative corporations were represented with magnificent pavilions designed by Walt Disney Imagineering. The Bell System, IBM, Eastman Kodak, General Motors, Chrysler, Ford, and General Electric were there. A stainless steel model of the world, 12 stories tall and canted at a 23.5-degree angle, was the centerpiece of the square mile (2.6 km²) park. About 51 million people visited the fair in its two summer seasons.

The General Electric pavilion, sitting next to the Clairol and Chunky Candy exhibits, was an enormous, inverted bowl, containing several interesting displays. The Carousel of Progress, which was moved to Walt Disney World in Florida after the fair, used audio-animatronic figures to demonstrate the changes in home life having taken

electrical power, this process would be repeated, collecting the energy released in each mini-explosion.

Experiments in this direction began in about 1965 at the Lawrence Livermore National Laboratory in Livermore, California. Although the primary mission of this laboratory had been hydrogen bomb development, within a decade its major funding would become ICF research, using the world's largest, most powerful lasers. The first attempt was the "4 Pi Laser." Twelve ruby-rod lasers were focused on a glass globe, eight inches (20 cm) in diameter, containing hydrogen. All 12 were fired at once, hitting the globe from all directions at once. Nothing near fusion conditions was achieved. Obviously, larger lasers would be necessary.

In 1972, the "Long Path Laser" experiment was constructed at Lawrence Livermore, using a newly developed type of high-power laser. It place in the 20th century. The Mammoth Sky-Dome Spectacular presented the struggle to harness atomic energy on a very large projection screen, and the Fascinating Medallion City showed how everything in the entire country could run on electricity. Of special interest was the Fusion on Earth exhibit.

In the center of the building was a large glass dome, housing a ZETA fusion reactor. It took half an hour to build up sufficient charge in a bank of electrical capacitors, hidden beneath the floor. On the hour and the half-hour, a voice would come over the speaker announcing an impending fusion reaction. Everyone's attention was directed to the glass dome as a hush fell over the multitude and movement stopped. There was a loud bang and a blue flash from the dome. That was it. On the wall, a digital display would suddenly start counting up, eventually indicating dozens of increments added to the existing number on the display. The total was in the millions. This was the number of neutrons detected from under the glass in all the thousands of times the machine had been set off for visitors, and it indirectly implied the number of fusion reactions created. This was the state of fusion power in 1964. Power production had yet to be achieved, and visitors to the General Electric pavilion were reminded that it might take another 30 years before usable electricity could be generated using this technology.

More than 7.4 million people saw the fusion reaction in 1964. After the summer season in 1965, the fair was dismantled. The stainless steel Unisphere globe remains to this day, as well as the concrete avenues and the remains of the New York City pavilion. Fusion power is still estimated to be 30 years in the future.

used neodymium glass rods to further amplify light from already high-powered pulse lasers. It was able to deliver 50 joules of heat and pressure into a fusion target at infrared frequencies for about 10 nanoseconds, or ten-billionths of a second. The targets were tiny beads of plastic, in which deuterium was substituted for hydrogen in the hydrocarbon material. The beam quality was poor, and fusion was not observed.

In 1975, the problem of the beams was addressed with a new laser, the "Cyclops." It used spatial filtering, which amounts to shooting the laser beam through a telescope, backwards, so that its beam is focused down to a pinpoint. This work led to the massive Shiva experiment, using 20 Cyclops-type lasers all focused to a fusion point. Shiva was expensive and huge, costing $25 million and occupying a building with a floor the size of a football field. Although it was able to deliver 10,200 joules to the

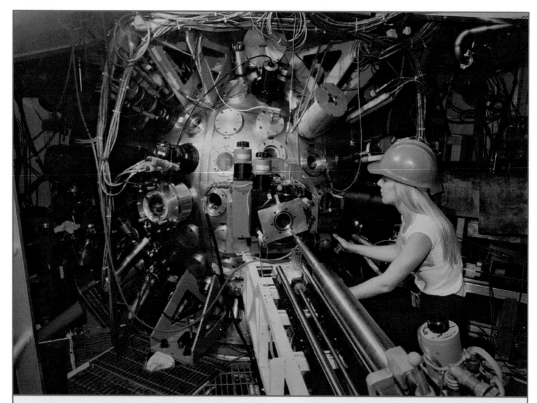

The Shiva target chamber at the Lawrence Livermore National Laboratory, California *(Lawrence Livermore National Laboratory)*

fusion target, its performance was disappointing, as it never approached ignition conditions. It did, however, prove that infrared light could not be expected to produce fusion. Shorter wavelengths would be necessary.

In 1984, a larger experiment named NOVA was built, this time using 10 ultraviolet light lasers with the intention of achieving one ignition, in which a fusion chain reaction is started. Useful power would be generated for a few nanoseconds, proving the concept. The setup could deliver more than 40,000 joules of energy to its target. At this unprecedented power level stability problems arose in the plasma produced by the short pulse of intense light. Fusion was achieved, but no significant power production was measured.

To address this instability problem, a larger, more precisely configured laser experiment was needed. In 1990, it was decided to build the *National*

Ignition Facility (NIF) at Lawrence Livermore, with the intent to achieve fusion ignition. This project is still underway.

The path to fusion has proven to be long and treacherous. The technology and the science are not simple, and there are problems that have not even been discovered yet. The next chapter will go into further depth explaining the twisted road of fusion research.

3 Magnetic Confinement Devices

The purpose of a fusion power reactor is to generate heat using the fusion of hydrogen isotope nuclei, two at a time, into a single nucleus of a heavier element. If this two-to-one reaction can be repeated billions of times per second, then significant energy is released and the heat can then be converted into electrical power by using it to create steam. The steam is used to drive a turbine, which drives an electrical generator in turn. To accomplish this purpose, the fusion reactor must create an environment in which these hydrogen isotope nuclei are heated to temperatures and compressed to densities that are much higher than those achieved at the center of the Sun. This is a challenging set of requirements, to say the least.

At the required temperatures of hundreds of millions of degrees, the hydrogen gas, or the fusion fuel, becomes ionized into a plasma. Plasma is a state of matter in which the normal conditions, whereby negatively charged electrons are found orbiting around positively charged nuclei, are not in effect. The nuclei are stripped free of all orbiting electrons, and electrons and nuclei exist together in a highly fluid soup, with the positively charged and the negatively charged components taking negligible notice of each other.

For a fusion to occur, two positively charged nuclei must become close enough together for a short-ranged strong nuclear binding force to stick them together. This coming together is always strongly opposed by Coulomb repulsion, or the tendency for like electrical charges to repel each other. The nuclear binding force can overcome the Coulomb force, but only at distances closer than about 1.7×10^{-15} meters (1.7 fm). Given this restriction, at the temperatures and pressures, or inter-nucleus proximities, that can be achieved in a compressed plasma, fusion is possible. At extremely high temperature the particles in plasma, electrons and nuclei, are traveling at near-light speed, and there are plenty of nearby particles to meet in a head-on collision. Every now and then, two hydrogen isotope nuclei will hit squarely going in opposite directions and stick together and, with the construction of a new nucleus, excess energy is released. If a volume of just a few cubic yards or meters of compressed hydrogen plasma can be kept together continuously under these conditions, then usable power is generated in a compact device that leaves behind no radioactive waste.

There are now 10 possible fusion reactions regarded as potentially useful in full-scale fusion power reactors using elements as heavy as lithium, shown in the table on page 48. The deuterium-tritium reaction is particularly attractive, because it requires the lowest ignition temperature, about 40 million degrees.

There is one basic problem with creating fusion in plasma. There is no vessel, no substance, and no container in which plasma can be kept. Such a container, necessary to localize a highly compressed substance, must have solid walls. Solid walls conduct heat, and if the plasma touches anything solid, it cools instantly. Cooled plasma is no longer plasma. Touching a wall destroys the fusion conditions instantly.

It is, however, possible to create a virtual container for plasma by exploiting the relationships between electrical charge and magnetism. A moving magnetic field creates an electrical field, just as a moving electrical field produces a magnetic field. Moreover, a static, or stationary, magnetic field will impose a force on an electrically charged particle. The force is in a specific direction, depending on the orientation of the magnet, and of a magnitude based on the strength of the magnet. This phenomenon, called the Lorentz force, can be used to manipulate charged particles in predictable ways, and this is exactly what is needed to not only contain plasma, but to even compress it without needing to touch it. The plasma is a fluid in which every particle has an electrical charge. It is a perfect substance for manipulation using magnetic force. It can be levitated, formed into a ball, squeezed, stretched, or made to spin at will.

This unique property of plasma, in which it can be controlled by a static magnetic field, was seen early in fusion research as the logical way

POSSIBLE REACTIONS FOR FUSION POWER PRODUCTION

Deuterium	+	Deuterium	→	Helium-3	+			Neutron	+ 3.2 MeV
Deuterium	+	Deuterium	→	Tritium	+			Proton	+ 4.0 MeV
Deuterium	+	Tritium	→	Helium-4	+			Neutron	+ 17.6 MeV
Deuterium	+	Helium-3	→	Helium-4	+			Proton	+ 18.3 MeV
Lithium-6	+	Proton	→	Helium-3	+	Helium-4			+ 4.0 MeV
Lithium-6	+	Helium-3	→	Helium-4	+	Helium-4	+	Proton	+ 16.9 MeV
Lithium-6	+	Deuterium	→	Lithium-7	+			Proton	+ 5.0 MeV
Lithium-6	+	Deuterium	→	Helium-3	+	Helium-4	+	Proton	+ 2.6 MeV
Lithium-6	+	Deuterium	→	Helium-4	+	Helium-4			+ 22.4 MeV
Lithium-7	+	Proton	→	Helium-4	+	Helium-4			+ 17.5 MeV

Note: This is a list of possible fusions of nuclei that could be useful for generating power. The first shown is a fusion of a deuterium nucleus with another deuterium nucleus, resulting in a helium-3 nucleus, a free neutron, and 3.2 MeV of usable energy. An isotope nucleus as heavy as lithium-7 could be used for fusion power.

by which it can be contained without a container and compressed without a compressor. Plasma could be created from fusion fuel and kept in a fusing environment by simply suspending it in a specifically designed set of magnets. The first experiments in the early 1950s involved hydrogen gas contained in a glass tube with a coil of magnet wire at each end. A high voltage electrical current passing though the gas would strip off the electrons and convert it to a glowing plasma. Suddenly energizing the two electromagnets compressed the plasma, heating it to extreme temperatures and keeping it away from the walls of the tube, approaching the conditions necessary for fusion.

There was one problem. This "magnetic mirror" device did not contain the plasma perfectly. No matter how powerful the end-magnets were made and how tightly the plasma was squeezed on the ends, particles still escaped out the ends and prevented the full conditions for fusion. Scientists and engineers would wrestle with this problem for the next 60 years. The quest for fusion in magnetically contained plasma continues to this day, and there are tantalizing designs that seem close to a solution.

THE TOKAMAK: A TOROIDAL CHAMBER IN MAGNETIC COILS

The end-leakage from a straight tube with magnets on the end can be eliminated by curling the tube back on itself, making a hollow donut, or toroidal chamber. If there is no end to the tube, then there can be no leakage from the end. After some preliminary results had been obtained using straight tubing, this corrective modification occurred to several independently operating research groups.

After the surprise announcement of successful laboratory fusion from Argentina in 1951, as detailed in the last chapter, physicists in the Soviet Union working at the Kurchatov Institute in Moscow devised and in complete secrecy built a toroidal fusion chamber named the tokamak. This type of device, using two known methods of magnetic containment in tandem, became the most widely copied fusion reactor in history. More than 200 tokamaks have been built and tested around the world, and there are presently 21 tokamak units in operation. The first tokamak built outside Russia, after disclosure of the design in 1963, was the LT-1 at Australian National University in Canberra, Australia. The latest experimental tokamak built is the KSTAR in Daejon, South Korea.

A tokamak consists of a sealed metal tank, shaped like a donut. The smallest examples will fit on a desktop, and the largest require two-story buildings. The tank or chamber is emptied of all air using a vacuum pump and is then back-filled with hydrogen, the hydrogen isotope deuterium, or a mixture of deuterium and the isotope tritium. An electric current is fed through the gas in the chamber. This both reduces it to plasma and causes a poloidal magnetic field, or a self-induced magnetic field in the plasma itself, causing it to compress away from the chamber walls and bunch up in the circle that defines the center of the donut-shaped void.

This poloidal magnetic field is important for creating the fusion conditions, but it is insufficient. It is uncontrollable and imprecise, and more compression and higher temperature are required. For enhanced magnetism, 32 D-shaped coils of wire are stationed around the torus, looping around the chamber with the straight sections of the D's facing the center hole in the donut. This externally imposed magnetic field intensifies the compression effect in the chamber. By shaping the coils in a specific way and controlling the electrical current that powers the magnets, the configuration of the resulting field in the torus can be tuned or formed so as to give an optimum fusion environment.

A third source of bulk magnetism for the tokamak is provided by at least one large transformer core sitting in the donut hole. The transformer core is a continuous loop of magnetic iron, with one half extending completely through the hole in the torus, and the other half off to the side of the tokamak, wrapped with wire, forming an electromagnet. Magnetism produced by current through the wire follows the circuit of iron, and it provides a powerful third source of magnetic force to the fusion device.

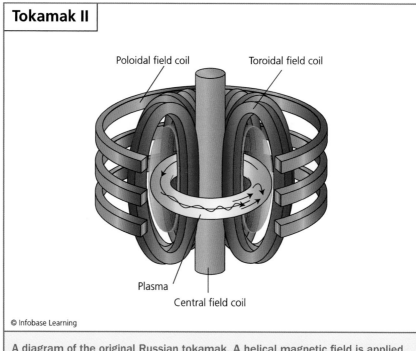

Tokamak II

Poloidal field coil

Toroidal field coil

Plasma

Central field coil

© Infobase Learning

A diagram of the original Russian tokamak. A helical magnetic field is applied to the torus by the iron rod in the center. The two additional magnetic fields are provided by coils of wire looping around the chamber in different directions.

As tokamak designs spread across the scientific world and the size of the toroidal plasma chambers increased, more intense magnetic fields were required in efforts to achieve chain reaction fusion. In 1988, the large T-15 tokamak at the Kurchatov Institute was completed. Bulk magnetism through the iron cores was provided by superconducting magnets, or magnets wound with wire having zero resistance to electricity.

The principle of superconductivity has been known for a hundred years. All normal metal used in electrical wiring, even copper, has a measurable resistance to electrical current. This resistance limits the amount of current that can be conducted through the metal, and this puts a limit on the degree of magnetism that can be attained using electrical wire wrapped around an iron core. In certain metals and special alloys, this resistance disappears at extremely low temperature, within a few degrees of *absolute zero.* Absolute zero is 459.67°F (273.15°C) below zero, and in this condition of absolute cold, also known as zero degrees Kelvin, all particle motion slows to a minimum. The magnets in T-15 were wound with niobium-titanium filaments, with copper added to provide mechanical strength and a safety path for electricity should the superconductivity fail. The magnets were cooled down using liquid helium at 4.2 Kelvins, or –452°F (–269°C). Using the supercooled niobium-titanium material, a magnetic field strength of 150,000 gauss (15 teslas) was possible—more than 10 times what could be attained using only copper wire. Since the successful use of superconducting magnets in T-15, this technology has been incorporated into every magnetic containment experiment built in institutes around the world. T-15 was shut down in 2005, due to a lack of funds, after proving the utility of superconducting magnets.

The first tokamak at the Kurchatov Institute was named T-1. Although early experiments with this configuration proved encouraging, a sustained fusion of hydrogen isotopes was not achieved. A minimum requirement was a higher temperature in the plasma, and subsequent experimental tokamak setups have been built with this goal.

Plasma temperature was difficult to determine. Any probe or sensor in the plasma chamber would not work, as the slightest intrusion would disturb the geometry of the inside of the torus and prevent magnetic containment. It was possible to have glass viewing ports in the chamber, but all the scientists could do was watch the plasma light up and writhe in the torus, without revealing the actual temperature. Temperatures were determined indirectly, and these methods were open to debate and dismissal. The 1968 announcement of a temperature of 20 million degrees in the Soviet T-4 tokamak was met with some skepticism, as it was hotter than any plasma reaction achieved in the United States or Great Britain.

British scientists had recently perfected a new, nonintrusive method to determine plasma temperature, and they were invited to measure the T-4 tokamak at the Kurchatov Institute in an unprecedented gesture of fusion research cooperation. The method employed a laser beam, directed through a transparent viewing port on the side of the toroidal fusion chamber.

The measurement is always referred to as temperature, but what is really being measured is the average speed of individual particles as they ricochet madly off each other in the hot plasma soup. Temperature is directly proportional to this speed, and the higher the speed, the more likely are fusion-causing collisions between particles. The laser light is a unique form of visible or near-visible radiation, in that it is composed of photons of a single frequency or wavelength forming a tight, nondispersive beam. The British scientists used a red laser beam, pointed at the center of the mass of plasma. Reflections off the plasma were collected with a photospectrometer through the same viewing port. The spectrum of the returning laser beam was examined to determine the extent to which the single frequency of the incoming beam had been spread out into frequencies greater and lesser than the original red laser. The beam of photons struck plasma particles and reflected back into the spectrometer. Those photons that hit oncoming particles came back with shorter wavelengths and higher frequencies. Those that struck outgoing plasma particles returned with lower frequencies and longer wavelengths. It is the Doppler effect, or the same principle that makes a police siren sound a higher note when it is approaching and lower when it is receding. The faster the police car is moving, the greater the spread between the frequencies of the coming and going sounds. In like manner, the extent of the spectral spread from the plasma indicates the speed or the temperature in the torus. The Soviets were correct. Their tokamak was achieving a momentary temperature of 20 million degrees.

For a fusion chain reaction in a plasma chamber, the temperature would have to be increased by a factor of 10. A new plasma-heating technique, high-power neutral beam injection, was planned for the $14 million Princeton Large Torus, or the PLT, at Princeton University in 1975. At the time, it was the largest tokamak in the world, weighing 150 tons (136 mt). The volume of the torus was 221 cubic feet (6.26 m^3). This ambitious fusion project was to use beams of complete hydrogen atoms having no net electrical charge, flung at high speed into the magnetically controlled plasma. These externally introduced particles would be free to slam into plasma particles, exchanging momentum and imparting an increased speed to the struck ions.

It seemed a straightforward plan, but implementation was difficult and time consuming. Four neutral beam injectors were planned, with a combined power of 2 million watts. It took three years to install the neutral beams and resolve technical problems, but the wait was worth it. The PLT was able to reach 65 million degrees in the plasma with the injectors working at full power. With further work, the plasma temperature reached a new record: 82 million degrees. The goal of fusion research at the time was to reach the break-even point, at which the energy required to initiate fusion was equal to the energy recoverable from the fusion. Actual energy production would then become another step-wise goal, but experiments were falling short of that initial target. Still, the encouraging performance of neutral particle injection gave hope, and it provided renewed credibility to the idea of fusion as a source of power.

By the 1980s, tokamak fusion reactors were producing significant but non-self-sustaining fusion events, and the amount of electricity required to power up the magnets and the external heat sources had grown along with the recoverable power from fusion. Although still not at the break-even level, the fusion action was sufficiently powerful to melt the inside wall of a torus and disrupt superconducting cooling in the magnets. The desired product of fusion is released energy, and a recoverable component of this energy is freed neutrons. The neutrons released by fusion have no electrical charge and are not constrained by the magnetic fields in the torus nor by the physical walls and support structures of the machine. They diffuse out of the plasma chamber, exchanging momentum with every atom they run into and heating up everything in the process. In a fusion power plant, this is an important feature, as it is a method of transferring energy to a working fluid, such as water turning to steam. In an experimental reactor, however, it is an engineering problem.

As tokamak fusion experiments increased in sophistication, active cooling became a design consideration. Plasma chambers were lined with ceramic plates for resistance to melting at high temperatures, and the outer surfaces of the tokamaks were cooled cryogenically, that is, with liquefied gases. Liquid nitrogen has been used successfully as a coolant, at –321°F (–196°C), and even liquid helium at near absolute zero has been used. These cryo-cooling systems increase the cost and complexity of tokamak experiments, while making the quest for a break-even state possible.

In 1980, the Princeton Plasma Laboratory built the Tokamak Fusion Test Reactor, or TFTR, hoping to achieve the break-even point using larger, more powerful neutral particle injection. This toroidal chamber plus the magnets used in this ambitious series of experiments weighed 700 tons (635 metric tonnes) and stood 30 feet (9 m) tall. There was room inside the torus for 10 people to stand. The budget for this project was $314 million, promised by the U.S. Department of Energy, but by the time TFTR was complete and running, the cost to build it had ballooned to almost $1 billion. The power bill for turning it on and running experiments was $20,000 per day and 1,200 people were on the payroll.

Princeton's TFTR project was finally shut down in 1997. It had been in direct competition with the Joint European Torus, or JET, in Culham, Oxfordshire, England, for becoming the first fusion reactor to achieve break-even conditions. Although both machines were able to reach temperatures of hundreds of millions of degrees using neutral particle injection, neither was able to approach the goal. JET remains in operation.

JET uses two methods of superheating the fusion plasma: neutral beam injection and radio frequency heating. This second method of heating something without touching it is known to anyone who has warmed a leftover casserole using a microwave oven. The oven is a resonant chamber, made to dimensions corresponding to the wavelength of microwaves introduced through a slit in the side from a magnetron oscillator tube. An electromagnetic standing wave in the oven excites water molecules in food, causing the temperature to rise as the water molecules become agitated and begin to bounce. A similar effect can be used on plasma particles, by choosing a microwave wavelength that will divide evenly into dimensions of the toroid fusion chamber. The JET has used 15 megawatts of radio-frequency power added to 23 megawatts of neutral particle injection in attempts to reach break-even conditions.

As tokamak design evolved over 30 years of experimentation, the fusion chambers were built larger with each increment, and the plasma volumes grew. More plasma requires more external heating to reach fusion temperature. Ion cyclotron resonance heating and electron cyclotron resonance heating were added to the available methods of increasing the plasma temperature. The cyclotron, invented by Ernest Lawrence (1901–58) at the University of California, Berkeley, is a powerful charged-particle accelerator using microwaves. In this case, the extremely high frequency of microwave radiation, in the billions of hertz, is used to electrically influence the particles. Each particle, either charged positive or negative, floating in a vacuum chamber, is alternately attracted and repelled by the oscillating electric field at the end of a wave-guide conducting microwaves from an oscillator. Back and forth the particles are drawn, billions of times per second. They cannot move in straight paths between the electrodes, because a powerful, static magnetic field crosses through, running up and down or perpendicular to the path between the electrodes.

The magnetic field forces the charged particles to move in a circle. Faster and faster they travel, and as their speed increases, the circle spirals out, with a larger and larger diameter. Finally, the particles are moving at near-light speed, and they leave the cyclotron through a slit in one of the electrodes. Positively charged protons, deuterons, or tritons and negatively charged electrons can be thrown at very high energy into a fusion chamber to increase the temperature of the plasma.

One of the largest magnetic confinement experiments in the world, the *Large Helical Device* in Toki, Gifu, Japan, directs every available external heating method into its plasma stream. It uses ion cyclotron resonance, electron cyclotron resonance, radio frequency heating, and neutral beam injection all at once, but it is not a tokamak. It is the alternative to the tokamak, a completely different device named the stellarator.

THE STELLARATOR: A TWISTED TOKAMAK

The tokamak, in which a stream of compressed, heated hydrogen plasma is bent into a circle, was a logical solution to the problem of plasma leakage from a linear fusion reactor. Surrounded completely by a strong magnetic field, there is simply no opening for escape. This magnetic cage would turn out to be less than perfect as the complex dynamics of plasma capture became known, but in theory the containment strategy was sound.

Even as the first tokamak was being assembled in secret, a flaw was noticed. The donut-shaped plasma chamber had occurred to others, as early as 1938, and it had occurred to the plasma physics team at Princeton

University that was assembled to build a fusion reactor in 1951. The flaw in this design was the fact that the inner radius of a donut or torus-shaped chamber is smaller than the outer radius. This means that any magnetic field created by coils of wire wrapped around the torus will be lopsided, or asymmetrical. The wire coils will be closer together on the inside than on the outside, so the resulting magnetic field will be stronger on the inside. Plasma established at the center line of the chamber is therefore pushed to the outside, and this is not an advantageous configuration.

There was an elegant solution to this problem, and it would become known as the twisted tokamak, or the stellarator. Instead of making one simple circle, the plasma in the stellarator followed a figure eight, crossing over at the center of an elongated plasma chamber. It worked like a slot-car racing track. If confined to a simple oval track, slot cars, which are forced to run in nonintersecting paths, would have unequal distances to travel around the circuit, and the cars on the inside would have a clear advantage over cars on the outside. Slot cars travel over a figure eight, so that the cars on the inside in half the track become cars on the outside on the other half of the track, and the total distances run by the cars are equal. So it was with the plasma in the stellarator. The volume of plasma, as it orbited around inside the chamber, pushed by the magnetic fields, would be crowded to the outside of the curved space by the uneven magnetic field. However, on an individual basis a plasma particle would find itself on the outside of the curve only half the time, and the other half of the time it would be on the inside of the curve as it moved between the two loops in the figure eight. On average, a given particle of plasma would experience the magnetic field at the center of the volume.

The stellarator fusion reactor design relies only on externally created magnetic fields to contain and compress the hydrogen plasma. The magnetism to compress the plasma is created by coils of wire wrapped around the cylindrical plasma chamber. Coils are spaced at equal intervals around the figure eight configuration, and each is activated by a constant electrical current. The first stellarator figure eight plasma chamber, as built in secret at Princeton University, was made of blown glass and was suspended by wires above a wooden table. Problems with the design became evident after initial testing.

The figure eight idea would have worked as planned if it were possible to build a flat figure eight. For practical reasons, as the two sections of

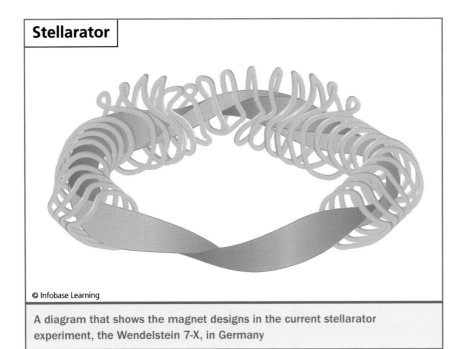

Stellarator

© Infobase Learning

A diagram that shows the magnet designs in the current stellarator experiment, the Wendelstein 7-X, in Germany

tubing passed each other at the center, in an X-configuration, they had to be at different heights, with one passing over the other. This necessity caused the curved sections at the ends to be tilted slightly to allow the two sections to miss each other in the middle. This perturbation prevented the plasma particles from being symmetrically exposed to the outer and inner walls of the chamber. Plasma particles turned out to be sensitive to the slightest irregularity in the magnetic containment field, and this included asymmetry in the usage of the magnetism.

Many different, oddly shaped plasma chamber configurations were tried to address this problem, including a peanut-shaped tube instead of a figure eight. The convex bends in the two ends of the peanut were compensated for by the concave bends at the middle, with the plasma spinning around inside. A more stable plasma could not be fully achieved with these mechanical solutions.

After decades of research, variations on the figure eight plasma chamber were condensed into a torus, or donut-shaped, tube. However, this configuration, named the torsatron, was not a tokamak or a version of the tokamak. It was still a type of stellarator. The plasma ring inside the torus, compressed

Princeton Plasma Physics Laboratory technician works on a National Compact Stellarator Experiment magnet coil *(Princeton Plasma Physics Laboratory)*

away from the walls of the chamber by magnetic field coils, was spun or rotated by a second set of external magnetic wiring. Under the circular coils, a second set of magnet wires was laid perpendicular to the compression coils. These wires were formed into a gradual helix, winding around the fusion chamber and causing the plasma particles to run in spiral paths around the donut. Given this new mode of movement, a plasma particle spent as much time near the outside track in the chamber as it spent near the inside track. Averaged over time, the path of any plasma particle was in the optimal place, at the center of the compressed ring. This design principle showed promise beginning in the 1980s.

Although the tokamak remains the preferred reactor configuration for most high-budget plasma fusion projects, there is still hope that sustained, break-even fusion may be attained using a variation of the stellarator. Although the National Compact Stellarator Experiment at the Princeton Plasma Physics Laboratory was cancelled in 2008, the Large Helical Device in Japan is still being used to refine plasma fusion techniques.

THE JOINT EUROPEAN TORUS REACTOR

There are literally hundreds of tokamak reactors built worldwide, and accounts of all cannot be included here, but a few examples have proven successful in stepping the science of plasma physics and magnetic fusion containment forward. A fine example is the Joint European Torus (JET Program), built on an old airfield in England starting in 1977.

The JET Program was conceived by a council of scientists of the European Atomic Energy Community (EURATOM) in 1970. The goal was a robust fission power project, to be jointly owned and paid for by the participating countries of western Europe and the United Kingdom. After the necessary legal framework was developed, design work on a large tokamak reactor began in 1973. The site at Culham, Oxfordshire, was chosen after much debate in 1977, and the Torus Hall building was completed in 1978. Components for the reactor were built all over Europe and transported to the site for installation. On June 25, 1983, the first plasma was achieved in the completed JET reactor. The project had been completed on time and on budget.

The power surges for the magnets on JET are beyond what the electrical distribution wires in Oxfordshire could provide, so two flywheel generators were built to take the load for one minute as the reactor sustains a

THE PERHAPSATRON

Since the beginning in 1951, funded fusion power research has been highly competitive. Part of the competition is to be the first team to achieve a workable fusion reactor, but primarily the brawl involves research money. Fusion experimentation, be it based on plasma or *laser* physics, is among the most expensive pursuits in science, and there is no bottomless source of funds. What used to be paid for with small pieces of federal budgets now requires consortia among several industrialized nations. Billions of dollars are required to begin a new fusion-reactor project.

In the early days, competition commenced immediately, as universities and national laboratories jumped for the same budget. In 1951, Princeton University and the Los Alamos National Laboratory were both eager to be the first to achieve a controlled thermonuclear action, even though all fusion research was a military secret at the time. Both Princeton and Los Alamos were funded by the Atomic Energy Commission. The name of this project was assigned because of a British scientist working on the hydrogen bomb project.

James Leslie Tuck (1910–80) was born in Manchester, England, and educated at Victoria University of Manchester. In 1937, he was offered an appointment at Oxford University as a research fellow. Here he had the privilege of working with Leó Szilárd (1898–1964), a brilliant refugee from Hungary, and the two let their imaginations fly into the frontiers of physics theory, including the concept of hydrogen fusion for power production.

Tuck and many other nuclear physicists wound up in New Mexico during World War II, working on the atomic bomb project at Los Alamos, and in 1950, he returned to the lab-

stable plasma. One flywheel provides the power for the 32 toroidal magnet coils and one for the inner, or poloidal, field coils. The two flywheels are slowly accelerated to high speed using electric motors connected to the power grid. During a plasma compression, the motors are turned off, and the flywheels turn two generators, using the energy stored in the spinning flywheels to provide the high surge power necessary for the magnets. The bulk magnetism core is powered from the normal electrical grid.

JET is a large experimental reactor. The donut-shaped fusion chamber is 19.42 feet (5.92 m) in diameter, with an 8.2-foot (2.5-m) donut oratory to work on the new, bigger weapon, the hydrogen bomb. His nickname became "Friar Tuck." His interest diverted off the bomb project onto what he considered an even more interesting problem, the nonexplosive use of fusion to generate power. He independently came up with a plan to build a toroidal plasma chamber, using electrical conduction to create a magnetic field and cause compression. Tuck gave his creation a modest name: the *Perhapsatron*. Funding was forthcoming, but another research effort in the nearby Hood Building had to be shut down. That project was "robbed to pay Friar Tuck," so the logical name for the fusion investigation was Project Sherwood. By 1952, Tuck had an operating Perhapsatron lighting up with purple plasma.

In 1958, the United Nations sponsored the International Atoms for Peace Conference at Lake Geneva, Switzerland. The United States was well represented with working models of the B-stellarator from Princeton and the Perhapsatron from Los Alamos set up in the conference hall. The Perhapsatron produced millions of neutrons per second, none of which was due to nuclear fusion. As was the case of other plasma experiments at the time, including the stellarator, the neutrons were released by high-voltage plasma hitting the walls of the fusion chamber. There was no actual fusion in either of the working models.

The Princeton team was disturbed by the sight of Tuck's impressive equipment at the conference, and back at their laboratory they spent a large portion of their year's budget proving that the Perhapsatron could not produce fusion. Their work was useful, as they were able to demonstrate that Tuck's device suffered from kink instability, or an inability to prevent the compressed plasma from writhing and crashing into the wall of the torus. Unfortunately, their stellarator suffered from similar plasma containment issues.

hole in the center. This chamber weighs 110 tons (100 metric tonnes) and is built primarily of carbon fiber composite material, lined with pure beryllium. The field magnet coils alone weigh 422 tons (384 metric tonnes), and the iron core in the bulk magnet weighs a hefty 3,080 tons (2,800 metric tonnes). It uses both neutral beam plasma heating and radio frequency heating, together requiring up to 38 million watts of power just to bring the hydrogen fuel up to fusion temperature. Using everything that was learned in the previous 20 years about plasma instabilities, the JET design can hold a plasma, compressed at the center ring

Inside the JET vacuum vessel with an image of the plasma superimposed *(EFDA JET)*

of the torus, for up to 60 seconds before it hits the wall of the chamber and extinguishes.

On November 9, 1991, JET accomplished the world's first controlled release of energy from a fusion reaction. In 1993, energy extraction was improved by rebuilding the inner fusion chamber with a plasma divertor mechanism. This modification was based on successful experiments at the *ASDEX* Upgrade, or the Axially Symmetric Divertor EXperiment, at the Max Planck Institute for Plasma Physics in Garching, Germany. Although about half the size of JET, the German reactor was able to divert contaminated plasma at the inner wall out of the compressed plasma core and therefore sustain the fusion for a longer time. Plasma was considered contaminated if it recaptured free electrons and became neutral. The divertor had a bonus function, in that it provided a method of capturing and measuring energy produced in the plasma stream.

Using this modification, in 1997, the JET reactor was able to produce 16 million watts of power from fusion. This remains a world record. Although JET was able to extract this amount of power from the plasma, it still required 24 million watts to heat the plasma to fusion temperature alone, not including the power necessary to sustain the magnetic fields. Beyond the reach of JET was the ignition point, the combination of magnetic force, plasma volume, and plasma temperature at which fusion will cause fusion, and not depend on power-hungry external equipment to

JET, looking toward Octant 2, taken in 1985 *(EFDA JET)*

sustain the heat production. Maintaining a plasma for one minute was also a milestone, much better than the one-millisecond plasma sustained by the pioneering tokamaks and stellarators, but it was still short of practicality. For realistic power production, a plasma would have to be maintained continuously for months.

Another milestone for JET was accomplished in 1998, when remote-handling equipment was installed for work inside the fusion chamber. The record-breaking power runs of the reactor in 1997, using a deuterium-tritium fuel mix, had created enough neutrons to activate materials in the machine to radioactive nuclides, and it became dangerous for unshielded humans to work inside the torus. This will be a safety consideration in future commercial fusion reactors, and the time was right to work out and test the designs of remote-operating robots.

The EURATOM partnership at JET began to break down as the British scientists, engineers, and technicians noticed that all the non-British staff from continental Europe were making twice their salaries. This was a system originally set up by EURATOM in 1973, but it began to seriously irritate the British workers. On their insistence, the practice was declared illegal, and in December 1999, the management of JET was taken over by United Kingdom Atomic Energy Authority (*UKAEA*).

JET remains in operation, undergoing modifications and experimenting, as a research facility should. It is making progress in improving the design of the future European ITER fusion reactor project by performing carefully planned tests of materials and equipment in a fusion environment.

TORE SUPRA AND OTHER ATTEMPTS TO IMPROVE MAGNETIC CONFINEMENT

In 1982, EURATOM decided to build its own large, high-performance tokamak reactor at the nuclear research center of Cadarache, Bouches-du-Rhône in Provence, France. The goal of this ambitious project was to achieve long plasma pulse durations, exceeding the 60-second run at the JET reactor. To accomplish this, a massive toroidal magnetism would be necessary, and this would require cryogenically cooled superconducting magnets. The new reactor was named Tore Supra, from the words torus and superconductor, and a race began with the Soviets for the world's first such fusion reactor. The T-15 tokamak at the Kurchatov Institute was being built in parallel with Tore Supra using superconducting magnets.

Tore Supra was completed and ready for operation in April 1988. It is not the largest tokamak in the world, with a fusion chamber diameter of only 5.9 feet (1.8 m), but the use of superconducting magnets is impressive. The mag-

An overall view of Tore Supra—a man is standing on a platform near the center *(CEA)*

netic field at the center of the ring of plasma is 45,000 gauss (4.5 T). The field in the JET reactor, which requires two flywheel generators plus line current to achieve, for a short period of time, peaks at the lesser value of 34,500 gauss (3.45 T). To accomplish this, the niobium-titanium magnet windings are cooled to 3.2°F (1.8°C) above absolute zero, using liquid helium-3 in super-fluid phase. It is difficult to hold this temperature on the magnet windings and keep the inside of the fusion chamber from melting when the plasma is kept at hundreds of thousands of degrees using 20 megawatts of electron cyclotron resonance and other methods. The inner wall of the chamber is cooled to 428°F (220°C) by pressurized water running through a hollow, reinforced space between inner and outer chamber walls.

Power in the form of heat is extracted from the plasma by a limiter ring, which resembles a metal catwalk along the bottom of the inside of the donut-shaped fusion chamber. On closer examination, the limiter is made of 576 hollow metal fingers running horizontally and all connected to a common water conduit. Together, the limiter ring and the inner-wall cooler can extract energy from the Tore Supra at a rate of up to 25 million watts. The ultimate goal of this experiment is to generate and recover 25 million watts of power for 1,000 seconds.

On April 12, 2003, the Tore Supra was able to maintain a stable plasma for 390 seconds, surpassing the previous record set by JET by a factor greater than six. However, this long-pulse operation is possible only at very low power. A stable plasma duration of 120 seconds is possible at a power of 3 million watts, 30 seconds at 7 million watts, and two seconds at 12 million watts. A pulse of 1,000 seconds at 25 million watts remains a targeted goal, but the advanced techniques developed in this experiment series for long-pulse plasma are seen as a significant step forward for future fusion reactor designs.

Other recent magnetic confinement fusion experiments include the *START* and MAST projects at Culham, and the *EAST* (Experimental Advanced Superconducting Tokamak) reactor in Hefei, China. The START, or Small Tight Aspect Ratio Tokamak, was a radical departure from the tokamaks being built in ever growing sizes. The START was almost perfectly spherical, 6.6 feet (2.0 m) in diameter, with a vertical hole through the middle, technically making it toroidal. The purpose of this investigation was to increase the plasma compression, and at that it was successful. The previous record for the β factor, or the ratio of the plasma pressure to the magnetic field pressure, was 12.6 percent, achieved on the DIII-D tokamak at General Atomics in San Diego, California. START reached a β of 40 percent, a significant milestone.

Another START milestone was the expense of the project. START was built with parts left over from previous tokamak projects, and the neutral beam injector was loaned to the Culham Science Center by the Oak Ridge National Laboratory in Tennessee. START was built on time and on a micro-budget, and operation started in January 1991. It remained in operation until midnight on March 31, 1998, when it achieved yet another milestone. START was the first tokamak to be given to another laboratory in another country. It was small enough to be transported, and it was donated, fully assembled, to the ENEA Research Laboratory in Frascati, Italy. The original British reactor has been reconfigured into an experiment named PROTO-SPHERA and is currently in use.

Using what they had learned from START, the British scientists at Culham built another spherical device named MAST, or the Mega Ampere Spherical Tokamak. It took about two years to build, and it was completed in December 1999. Impressed by the performance of the START reactor, both EURATOM and the British UKAEA funded this experiment. The MAST configuration is a departure from all previous designs. The fusion chamber looks like a room-sized paint can, with a magnet rod running down the center, top to bottom. Magnet coils located inside the chamber, instead of outside, create a vaguely spherical region inside.

Experiments conducted with the MAST reactor include tests in September 2007 of diamonds as possible lining material for fusion power reactors. Although this would seem an expensive component for use in a commercial power plant, material cost has seldom stood in the way of a good fusion power-plant concept. MAST remains in place at Culham, further testing materials and magnetic field configurations for future increments in magnetically confined plasma.

The Chinese Academy of Sciences, in the capital city of Anhui Province in Eastern China, Hefei, has entered the plasma physics community with a small tokamak, designated HR-7U. To the world outside China, it is known as the EAST. It is distinguished as the first tokamak to be built using superconducting niobium-titanium windings for both the toroidal and poloidal magnets.

Compared with other advanced, conventional tokamaks, the EAST is modestly sized. The outer diameter of the fusion chamber is 11.2 feet

(3.4 m). The magnetic field at the center of the plasma ring, produced by two tiers of superconducting coils, is 35,000 gauss (3.5 T). The plasma temperature is enhanced by 7.5 million watts of externally applied power. Construction of EAST was completed in March 2006, and its first plasma was created on September 28, 2006.

The first impressive milestone of EAST was the cost of building it, which was only ¥300 million, or about $37 million. A superconducting tokamak is an extremely complicated and expensive item, and this budget is about one-twentieth of the cost to build the same device in Europe or the United States.

With the EAST reactor, built with the latest knowledge refined from 50 years of incremental experimentation in tokamak design, the Chinese plasma physicists hope to do what the Tore Supra has tried to do: hold a stable plasma for 1,000 seconds. They may well succeed.

There is another, completely different path being taken to achieve self-sustaining ignition in hydrogen fusion. It is inertial confinement, and its challenges are every bit as daunting as those faced by the tokamak builders.

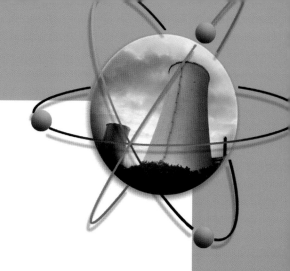

4 Inertial Confinement Devices

The purpose of magnetic confinement is to compress a plasma of hydrogen isotopes, cause it to start fusing, and keep this configuration stable, with the heat generated in the center of the machine maintaining a continuous state of fusion. The excess heat created by the fusion process can then be siphoned off and converted to electrical power. The goal is to sustain the fusion for months or even years in the donut-shaped chamber, replenishing the hydrogen and keeping the superhot plasma from melting the power plant.

What seems to be an elegant thought experiment turns out to be a great deal more difficult to achieve than it would have seemed. Perhaps this method of taming the process that powers the universe is going at it wrongly. On the low-volume scale, in which the plasma mass is much smaller than a star, the practical problem of maintaining the proper conditions for fusion has been challenging. There is, however, a completely different way of creating fusion and extracting energy from it. Instead of trying to hold a plasma for an indefinite time, assume that the fusion is a brief event, and design a power plant with that as a beginning assumption.

In fact, the most successful application of hydrogen fusion at this point is the hydrogen bomb, or the thermonuclear device. Intended as a last-resort weapon in superpower-to-superpower conflicts, the hydrogen bomb produces a very large spike of energy in milliseconds of hydrogen isotope fusion. The explosion is not very efficient, as most of the hydrogen is lost as the device rapidly destroys itself in a ball of glowing plasma, but for the amount of energy produced, the fuel and the associated equipment is inexpensive.

The only reason that the fusion process is kept together long enough for a significant amount of the hydrogen to experience fusion is an artifact of classical physics: inertia. Even with a hydrogen bomb explosion giving a big push, there are limits to how fast a hydrogen atom can get out of harm's way. Or, as Sir Isaac Newton (1643–1727) stated in his famous first law of motion, "An object that is at rest will tend to stay that way." The hydrogen atom will be moved away from the explosion, but at a finite rate. The inertia of the hydrogen, which is the lightest element in the universe with the least amount of inertia, is sufficient to keep a hydrogen bomb together long enough to annihilate its target.

A typical large hydrogen bomb produces an explosion the equivalent of 10 megatons of TNT. This energy spike is 4×10^{16} joules. The entire electrical energy consumption for the United States is about 100 megatons per day, so if it were possible to set off 10 thermonuclear devices and recover the energy, it would power the entire country on fusion for 24 hours on a hot summer day.

If the energy spike of a thermonuclear device could be dialed down to a reasonable, manageable level, then this method of energy production could be possible. By setting off very small hydrogen bombs, one after another, energy could be recovered in a pulsed mode, instead of the continuous delivery from a magnetically confined plasma. The plasma from the hydrogen explosion would be confined only by its inertia.

The necessary conditions of heat and pressure for converting hydrogen gas to hydrogen plasma and causing it to fuse are established in a hydrogen bomb by an adjacent atomic bomb, or a hypercritical fission device. This is not practical for miniaturized hydrogen explosions, as a fission bomb requires a specific mass of plutonium or uranium. A substitute was postulated for the initiating atomic bomb. Instead, a carefully directed laser beam could be used as the hydrogen explosion trigger device. The concept was proposed with a power plant design in the article, "Fusion by Laser: Experiments Indicate That Energy-Releasing Fusion Reactions Can Be Initiated and to Some Extent Controlled Without a Confining Magnetic Field by Focusing a Powerful Laser Pulse on a Frozen Pellet of Fuel," by Moshe J. Lubin and Arthur P. Fraas, in *Scientific American* (June 1971). In their power plant design, a small pellet

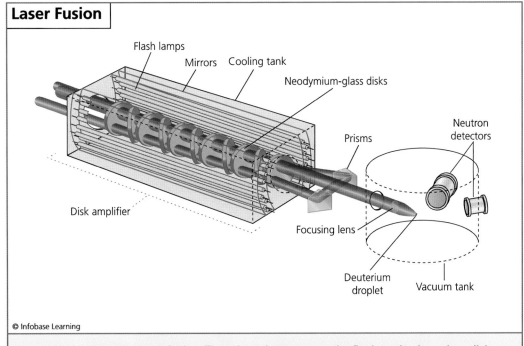

Laser Fusion

Flash lamps
Mirrors   Cooling tank
Neodymium-glass disks
Prisms
Neutron detectors
Disk amplifier
Focusing lens
Deuterium droplet   Vacuum tank

© Infobase Learning

The original concept for laser fusion: Three laser beams enter the final neodymium-glass disk amplifier at the upper right; the final beam is focused down onto a droplet of deuterium, which is expected to fuse under the pressure from the amplified light.

of frozen deuterium-tritium mixture is dropped into a spherical tank of swirling liquid lithium. When the pellet has reached the center of the tank, at the bottom of a vortex in the liquefied metal, a laser beam shoots down through the pellet insertion pipe, instantly heats the pellet to fusion temperature, and the resulting small thermonuclear explosion is absorbed by the lithium. The lithium, heated further by the explosion, is circulated around to a steam generator, and the steam is used to drive a turbogenerator in the usual manner.

This process is repeated, over and over, as quickly as is mechanically possible, and 16 lithium tanks operate in tandem in a typical power plant, generating electrical power without making highly radioactive fission products and using readily available fuel.

In practice, this process has been more difficult to realize than was initially thought.

Laser Fusion Reactor

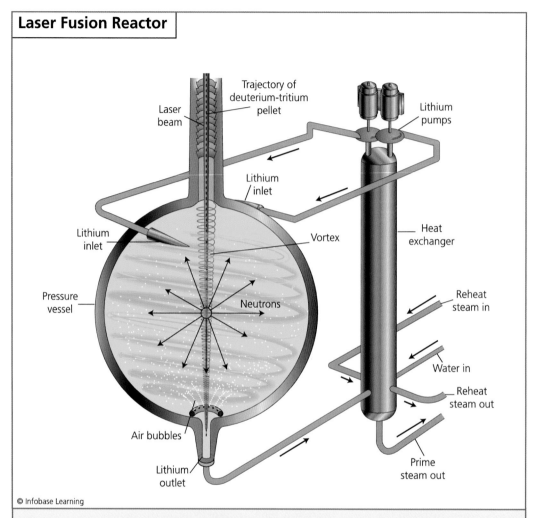

© Infobase Learning

Assuming that the laser-beam setup would cause deuterium to fuse, the energy from these fusions could be captured and used to run a turbo-generator using this setup. Frozen deuterium-tritium pellets are dropped one at a time into the pressure chamber, where they are fused with a laser beam. Heat from the fusion is absorbed by liquefied lithium metal, which is pumped through a heat exchanger to produce steam.

THE Z-PINCH DEVICE

Following World War II, British scientists began investigations into possible energy production using the nuclear processes fission and fusion. The newly formed Atomic Energy Research Establishment assumed possession of a surplus Royal Air Force base near the town of Harwell in Oxfordshire, and work began in the airplane hangars and related buildings. Early

A photo, taken on January 24, 1957, that shows David Goodall operating a model of the Z-pinch fusion reactor, with a pinched electrical discharge to show the action inside a glass torus, at Harwell, England (© *Hulton-Deutsch Collection/CORBIS*)

experiments involved hydrogen plasma contained in a glass tube. The tube was mounted vertically, and this corresponded to the Z-axis on a mathematical plot of three-dimensional data. The effect of the self-generated magnetic field in the vertical column of plasma was thus named the *Z-pinch*.

The pinching effect of a current running through a plasma is due to a previously known phenomenon, called the Lorentz force, discovered by the Dutch physicist Henrik Lorentz (1853–1928) in 1892. If two electrical wires are mounted parallel and both are carrying current running in the same direction, then the two wires will be pulled toward each other, due to the magnetic field generated by the moving current. Substitute a linear column of plasma for the wires, and the force of the magnetic field tends to compact the plasma along the main axis between the positive and the negative electrical terminals. Plasma is an ideal electrical conductor, having superconducting properties. It therefore seems possible to accomplish the effect of a stellarator, which uses an externally generated plasma-confining magnetic field, without using magnets.

Although it is technically a magnetic plasma confinement technique, the Z-pinch is considered to be an inertial confinement because no actual magnets are involved. Unlike magnetic confinement techniques, such as the stellarator or the tokamak, there is no opportunity to tweak and tune the magnetic field to try to hold the plasma for times longer than microseconds, so the Z-pinch is considered a pulsed fusion generator, and this is a primary characteristic of the inertial confinement strategies.

Early findings of success in fusing hydrogen isotopes were found to be erroneous in the large ZETA fusion reactor built at Harwell in 1954, and this letdown tended to steer funded fusion-reactor work toward the external magnet, torus-shaped experiment designs in the 1970s. The false fusion signals in this and several other apparently successful fusion experiments were neutrons created not by the plasma compression, but by another process. It is possible to create hydrogen-isotope fusion, and therefore neutron exhaust, by simply electrically accelerating a deuterium ion and crashing it into a stationary tritium ion. This is a common method of generating neutrons on demand, and it is used in applications ranging from industrial activation analysis machines to fission initiators in nuclear weapons. In the case of plasma experiments, stray tritium ions embedded in the inner wall of the fusion chamber are struck forcibly by deuterium ions when the plasma inevitably hits the wall, and this causes an encouraging spray of neutrons. The source of these neutrons turns out to have nothing to do with the hoped-for plasma compression and is instead due to the failure of the process. The deuterium-tritium fusion due to electrical acceleration is an easy but extremely inefficient process, with no hope of ever reaching a break-even state, in which as much power is used to create the fusion as can be recovered from it.

Still, in spite of the spectacular ZETA failure at Harwell, Z-pinch machines have been built and tested in the United States at Cornell University, Sandia National Laboratories, and the Nevada Terawatt Facility. There have been Z-pinch experiments conducted in Germany at Ruhr University, in the United Kingdom at Imperial College, in France at École Polytechnique, and at the Weizmann Institute of Science in Israel.

An odd embodiment of the Z-pinch principle is currently under study and development at Sandia National Laboratories in Albuquerque, New Mexico. Although its initial intent was not to seek fusion power, it could be a backdoor entrance to a workable power plant design. The purpose of the Z machine was originally as an inexpensive substitute for nuclear weapons testing.

In July 1974, the United States and the Soviet Union agreed to and signed the Treaty on the Limitation of Underground Nuclear Weapons Tests. It had already been agreed in 1963 that there would be no above-ground testing of nuclear weapons, and this agreement further stipulated that no weapons above an explosive yield of 150 kilotons ($6{\times}10^{14}$ joules) could be exploded. Thermonuclear weapons, or hydrogen bombs, typically deliver an explosion much larger than 150 kilotons, so further development of these high-yield devices came to a stop. A further Comprehensive Nuclear-Test-Ban Treaty of 1996 prohibits any nuclear weapons testing of any type. Although this treaty has been signed but not ratified, its specifications are effectively being followed, for budgetary reasons if not for the preservation of world peace.

However, these test bans did not diminish the need for data from nuclear weapons tests, and the Z machine, a Z-pinch device designed to test materials in extreme conditions, was built in secret. It is the largest X-ray generator in the world. As originally built, it could produce a quick pulse of 50 trillion watts of power. One pulse was a tenth of a microsecond long. Further development increased the output to 290 trillion watts. At this level, the Z machine produced 80 times the world's electrical power output for 0.07 microseconds.

The machine fires a very powerful discharge of electrical current, several tens of millions of amperes, into a cylindrical array of vertical, parallel tungsten wires. The electrical current vaporizes the wires instantly into a cylindrical curtain of tungsten plasma. The current is still on, and it passes through the now superconducting plasma, creating a Lorentz force that squeezes it into a thin, vertical shaft. The sudden, inward acceleration of the electrical field causes a pulse of X-rays of similar intensity to those produced in hydrogen bombs, and the resulting shock wave causes an extremely high pressure and temperature effect on a sample held in the center of the cylinder.

The small capsule in which a test sample is held is called the *hohlraum,* or hollow space, in German. An intense electromagnetic pulse of energy is a by-product of the Z machine, and this is also an effect of hydrogen bombs that is worthy of study. When the Z machine is fired, metal objects in the room surrounding it light up with a flashover, as high-voltage sparks are induced in anything that conducts electricity. Power feed lines are submerged in concentric chambers of 500,000 gallons (2 million l) of transformer oil and 600,000 gallons (2.3 million l) of deionized water to prevent cross-conductions.

It was found that the Z machine may have applications other than nuclear weapon simulations. It can, for example, accelerate small steel plates to a speed of 21 miles per second (34 k/sec), which is three times the Earth's escape velocity, theoretically flinging them into orbit around the Sun. It can compress water vapor to pressures of 120,000 atmospheres (12 GPa), creating a hyperdense hot ice. In 2006, the tungsten wires were replaced with thicker steel wires, and a record high temperature of 6.6 billion degrees F (3.7 billion K) was achieved at the hohlraum.

A logical application of such a device was hydrogen fusion, and, on April 7, 2003, Sandia Laboratories announced that it had fused deuterium in the Z machine hohlraum.

A high-voltage building up at the Z-pinch machine at the Sandia National Laboratories in New Mexico (Sandia National Laboratories; photographer, Randy Montoya)

The Z machine was dismantled in July 2006 for a $90 million upgrade to a power output of 350 trillion watts. Its new name was ZR, or the Z refurbished machine, and the work was completed by October 2007. The performance of the improved Z machine was impressive, and there are now plans for a ZN, or Z neutron machine, upgrade to a pilot fusion plant capable of a significant fusion shot every 100 seconds. The next planned step will be the ZIFE, or Z inertial fusion energy machine. If built, it will be the first true Z-pinch driven fusion power plant and could beat all other designs for the race to produce usable power from a fusion device. The extreme temperatures possible with the Z machines could also make it possible to fuse simple hydrogen with lithium or even with boron. These two fusion reactions do not produce neutrons, and this opens the possibility of a fusion reactor that produces no radiation and no radioactive by-products. A nagging problem with deuterium-tritium fusion reactions is that, although they do not produce heavy radioactive isotopes as does fission, they do produce free neutrons. Free neutrons activate surrounding structural materials into radioactive isotopes.

The Z-pinch fusion reactor idea is therefore not dead, despite some early disappointments, but another concept, laser fusion, is receiving even more attention and more funding.

SHIVA, NOVA, AND VERY LARGE EXPERIMENTS

In theory, deuterium-tritium fuel can be driven to fusion conditions using laser light, and this was seen as a possibility shortly after the first laser was built in 1960. Although the earliest experiments were not spectacular, it did seem possible to detect a neutron pulse when a ruby-rod laser was fired into deuterium-tritium gas. Over the next 50 years, laser fusion enterprise would grow in leaps from tabletop experiments to setups the size of football fields and budgets in the tens of billions of dollars. The reasons for this intense interest in inertial confinement fusion were not entirely due to a seemingly remote possibility of commercial power production. Behind the public show of fusion power by lasers were military purposes. An inertial confinement fusion device was an excellent way to study the details of how a thermonuclear weapon works in miniature, without causing a seismically detectable explosion. The institution that was given the most funding for this research by the Department of Energy was Lawrence Livermore National Laboratories, a facility that was built for the purpose of research and development of thermonuclear weapons, or hydrogen bombs.

A laser-driven inertial confinement reactor actually explodes a miniature hydrogen bomb, producing a scaled-down version of everything that

THE INVENTION OF THE LASER

One of the most important and influential inventions of the 20th century is the laser, or Light Amplification by Stimulated Emission of Radiation. The laser is a very precise light source, giving a parallel beam of light waves of one wavelength, all vibrating in phase. Lasers have been used in a wide variety of applications, ranging from a weapon that shoots down guided missiles to a handheld device used to single out interesting points on the screen during a Power Point presentation.

It is difficult to attribute the invention of the laser to one individual. It was an idea that seemed to reach a critical, explosive point in 1959, and there was no stopping it from making the transition from physics theory to engineering models making red spots on laboratory walls. The theoretical development began in 1916, with a paper published by Albert Einstein (1879–1955). Although he was philosophically opposed to the then new branch of physics called quantum mechanics, he had to give it its due, and his paper was foundational. It was known that when a photon of light reflects off an object, it is actually absorbed by the atom, causing an electron in orbit around the atom's nucleus to jump to a higher energy state. The higher state immediately decays, sending a new photon of the same energy or wavelength flying back out of the atom. What if, Einstein postulated, the higher energy state did not decay? If another photon then hit the same atom, the atom would let fly two photons of the same energy. A single quantum of light would be multiplied by the integer two. It would appear to the observer as if the reflection were twice as bright as the original light. The process was called stimulated emission of radiation, and the effect was light amplification.

This interesting effect went unapplied until a spring morning in 1951 on a park bench in Washington, D.C., when Charles H. Townes, a physicist from Columbia University, had an idea. He had been working on microwave radar since World War II, and it occurred to him that by exciting the electrons in atoms to a higher energy state using microwaves, he could achieve microwave amplification. He jotted down some quick notes and named his invention maser, or microwave amplification by stimulated emission of radiation. He and two graduate students built the first working maser in 1954, using ammonia gas as the amplifying medium.

happens in one of these high-yield weapons. The bomb is a tiny, frozen ball of a mixture of deuterium and tritium gases, typically the size of a pinhead and perfectly round. This pellet contains about 0.00035 ounces (10 mg) of fuel. An intense pulse of laser light hits it from all sides simultaneously.

Townes and his brother-in-law, Arthur Schawlow (1921–99), seeing further application of this principle, published a speculative paper in *Physical Review* in 1958 on an optical maser, or the amplification of light using a similar scheme. It was an interesting idea, but there were no known light-amplifying media to suggest.

In 1959, Townes organized a conference held at the Shawanga Lodge in the Catskill Mountains, New York, to discuss quantum electronics and the concept of an optical maser. Gordon Gould (1920–2005), a physics graduate student at Columbia, raised his hand in the meeting and suggested that the optical maser should be named laser. Townes and Schawlow rejected the idea out of hand. The many attendees, excited by this and other spirited discussions, emerged from the conference running to their laboratories to be the first to achieve light amplification. It was now a race, and the betting money was on Townes at Columbia or Schawlow at Bell Labs. Gould had been thinking quietly about his laser since 1957, and he had been keeping a dated notebook full of ideas.

After Townes's conference, Gould quit his studies at Columbia University and went to work at the Technical Research Group (TRG) in Melville, New York. He was able to convince his employer of the importance of the laser concept. TRG obtained a research grant from the Advanced Research Projects Agency (ARPA) and applied for a patent in Gould's name. The U.S. Patent Office turned down his application, and ARPA classified the project SECRET. Unfortunately for Gould, he was denied a security clearance because of his dabblings in communist activities as a student, and he could not work on his own project. Without Gould, the laser work at TRG made no progress.

The first working laser was powered up on May 16, 1960, by Theodore Maiman (1927–2007), using a rod-shaped synthetic ruby as the amplifying medium. Townes won the Nobel Prize in physics in 1964 for his work in stimulated emission of radiation, along with two Russians, Nikolay Bosov (1922–2001) and Alexander Prokhorov (1916–2002). Gordon Gould was awarded the patent for the laser in 1987 after three decades of legal battling. Although he had signed away 80 percent of any proceeds to pay for the legal costs, Gould eventually made millions of dollars in back payment of royalties from the well-established laser industry and was awarded 47 other patents for lasers and laser-related inventions.

Inertial Fusion

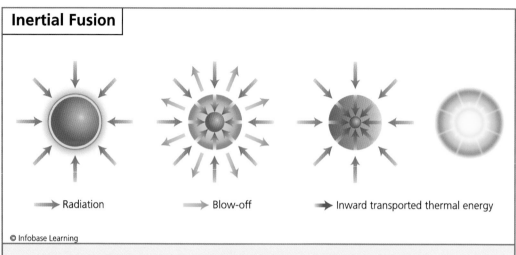

→ Radiation　　　　　　→ Blow-off　　　　　　→ Inward transported thermal energy

© Infobase Learning

In theory, radiation from all sides can compress a droplet of fusion fuel until it fuses. The surface of the droplet will blow off from the resulting energy release, but the explosion also points inward. This component of the energetic event compresses what is left of the droplet and produces a nearly complete fusion of the fuel.

The outer layer of the fuel pellet, heated suddenly, explodes. Material flies off the surface at high speed, but the inner surface of the exploding portion of the fuel also reacts, pushing inward. The outer shock wave dissipates as it moves outward, gaining size, but the inner shock wave becomes smaller and smaller as it travels in the opposite direction. The force concentrates as the radius of the shock wave diminishes, and it pushes the frozen fuel in the center together. The core of hydrogen isotopes quickly reaches the density of lead, and the temperature soars, to tens of millions of degrees.

Fusion begins, as deuterium and tritium atoms, forced close together in the high-density core, are induced to high-speed collisions by the extreme temperature. The first fusions release energy, and this increases the temperature and pressure in the remaining hydrogen, causing ignition, or fusion-initiated fusion. A chain reaction ensues, and all hydrogen isotopes become helium in the ball of plasma. If all of the fuel is consumed by fusion, then this pinhead-sized ball of hydrogen releases the same amount of energy as that produced by burning a barrel, or 43 gallons (159 L), of oil. If micro-bombs can be exploded at a rate of several per second, then excess energy from the fusion events can be drawn out of the process and converted to electricity. That is the goal of the power-production portion of the laser-driven inertial confinement research.

The first conceptual designs of laser fusion power reactors in 1970 assumed that one laser beam would strike the fuel pellet from one direction. The most powerful laser at the time could deliver about 1,000 joules of energy in a single pulse. It was predicted that a 100,000-joule laser would be required to produce significant fusion. Initial experiments at Lawrence Livermore indicated that not only would a larger laser be necessary, but that the pellet would have to be hit from all sides by a collapsing, spherical wave front.

In the 1970s, laser amplifiers were developed that would increase the power of a beam to the required level. In a laser amplifier, neodymium-glass disks receive energy transferred from arrays of xenon flashlamps, similar to but much larger than those used in digital cameras. A laser beam pulse entering a collection of 11 neodymium-glass disks as the lamps are flashing leaves the other end of the device amplified to a greater power. To reach the required power level required 100 such amplifiers, and to synchronize laser beams coming from multiple angles into a sphere meant that the pulse from one laser was optically divided up into separate beams, with each divided beam diverted into a neodymium-glass amplifier. Amplified beams were then sorted by complex mirror systems and each was assigned an angle to intersect the target pellet. All the beams were assembled into the desired spherical wave front in the center of the massive collection of amplifiers and mirrors.

Extreme precision was necessary in making each amplified laser beam. The beams had to converge at the target spot with an accuracy of plus or minus 0.00004 inches (1 μ). To ensure that all beams would arrive with this level of accuracy, it was important that all the beam-lines were exactly the same length.

The first large-scale laser fusion experiment was built at Lawrence Livermore National Laboratories in 1977 on a budget of $25 million. The project was named "Shiva," for the Hindu god having multiple arms resembling the many beam-paths of the divided laser pulse. The neodymium-glass amplifiers were able to deliver 10.2 joules of energy to the deuterium-tritium target with a pulse length of one nanosecond, or one-billionth of a second. The laser light was in the infrared spectrum, just out of range of visibility. Experiments started in 1978.

Although Shiva was a proof-of-concept experiment that was never intended to achieve full ignition fusion, initial results were disappointing. The ball of plasma created at the target site did not act as had been predicted, a problem that had plagued all other fusion experiments. A major finding was that infrared light was inappropriate for the desired effect. The next laser fusion reactor would use ultraviolet light.

A significant modification to the laser fusion scheme was tried using the Shiva setup. All previous attempts at fusion using lasers had used a direct mode of compressing the fuel, where laser light impinged directly on the hydrogen isotopes. If the target were considered to be a miniature hydrogen bomb, then that mode of compression was inappropriate. In a hydrogen bomb, the deuterium-tritium is compressed not by light, or heat, or even the explosive shock wave from the initiating atomic bomb, but from the first radiation to emerge from a fission explosion. That initial burst of energy is in the form of a wave front of X-rays, so powerful that they are able to compress the hydrogen isotopes into a fusion plasma. To achieve that same effect in the laser experiment, the deuterium-tritium was contained in a hohlraum, a cylinder of gold smaller than a pencil eraser. Hohlraums were built by making a plastic model, gold plating the model, and then burning out the plastic, leaving a hollow cylinder to contain the frozen deuterium-tritium droplet.

When the gold shell is struck by the laser beams from all sides, it flashes into non-fusible plasma. The sudden acceleration of the charged particles of gold creates a powerful X-ray wave front, with a large portion converging into the heated deuterium-tritium mixture. This indirect compression mode was an important advance in laser fusion, but no more than 100 billion neutrons could be produced per shot. This was far short of the ignition point, and there was no fusion chain reaction detected.

On January 24, 1980, a magnitude 5.5 earthquake hit Livermore, California. The building housing Shiva shook hard enough to shear off bolts and send the precisely positioned laser beams into disarray. Fortunately, the research team had already planned and designed the next experiment, an even larger array of laser amplifiers named Nova. Shiva was dismantled starting in 1981, but before the beam-paths and the target area were taken apart, the Walt Disney film *Tron* was filmed in the impressive building. Nova was built in 1984, at a cost of $174 million.

Nova was designed to achieve ignition and a fusion chain reaction in the hohlraum, and to accomplish this it was larger and more powerful than Shiva, using ultraviolet light and indirect compression. The laser pulse was divided into 10 amplified paths, each 600 feet (180 m) long, delivering 16 trillion watts to the target. The neodymium-glass amplifiers were inefficient at turning the flash-lamp light into amplified laser power,

This miniature star was created in the NOVA laser target chamber as 300 trillion watts of power hit a 0.5-millimeter target capsule containing deuterium-tritium fuel *(Lawrence Livermore National Laboratory)*

so it would require 100 times that much power into the system to make this laser beam. It would not be possible to recoup this power surge even if all the target hydrogen were fused, and the goal of a productive power plant using this technology was distant.

Early in the construction phase, an error was found in the design calculations. A design review was quickly convened, and it was confirmed that there was no way Nova was going to achieve ignition. Nevertheless, Nova was built and in December 1984 was switched on.

Nova was able to produce 100 times more neutrons per shot as Shiva, but the plasma problems multiplied accordingly. There were difficulties in synchronizing the 10 beam paths, and this led to hot spots on the target. The plasma ball became unstable during the compression and would not collapse uniformly. Still, the experiments were productive, and the understandings of inertial confinement fusion advanced, as well as plasma physics in general and even understanding of the evolution of galaxies and supernovae.

The NOVA laser bay at the Lawrence Livermore National Laboratory taken shortly after completion in 1984 (*Lawrence Livermore National Laboratory*)

A bigger, more powerful laser system would be necessary to achieve ignition, and a proposal for a $1 billion upgrade for Nova was sent to the Department of Energy in 1989. This budget seemed large to the project sponsor. When Lawrence Livermore added that the building then housing Nova, which covered the size of a football field, would have to be enlarged by a factor of three, interest in further funding began to wane. Nova was dismantled in 1999 to make way for a laser fusion reactor large enough to achieve the elusive ignition point. It would be named the National Ignition Facility (NIF).

The target chamber from Nova was lent to a team in France. Near Bordeaux, the French nuclear science directorate, Commissariat à Pénergieatomique (CEA), is building a very large laser fusion reactor. It is to be named *Laser Mégajoule (LMJ)*.

This inertial confinement device is expected to deliver 1.8 million joules of ultraviolet light into the hohlraum to produce fusion at the ignition point using indirect drive. LMJ will use an impressive array of 240 beam lines, grouped into eight clusters of 30 lines, each with two neodymium-glass amplifiers mounted in series. Research into fusion power production is an excellent cover story, but as is the case of many laser fusion devices, the main objective is to refine fusion calculations for nuclear weapons design. Construction of LMJ is expected to be completed in 2012.

Other large laser-fusion devices built outside Livermore include the LULI2000, the NANO2000, and the PICO2000 at École Polytechnique in France. These lasers are being used for more specific investigations, such as the coupling of nanosecond and picosecond laser pulses and the focusing of beams into spots of extremely small diameter. Findings from these experiments are then being used in the design of the next step, the HiPER, or the High Power Laser Energy Research facility. HiPER is to be built by a consortium of countries in the European Union, and it will be the first inertial confinement device to use the fast ignition approach to generating fusion.

At the completion of the NOVA project, laser fusion seemed to have reached an end-point at which the only way to go forward was to triple the size of the laser. Problems were once more with the plasma ball. Hot plasma would tend to mix with cool plasma in unpredictable ways, and this would quench the fusion before the ignition point. Increasing the laser energy was a brute force solution to this problem. A new design was proposed, in which two laser pulses were trained at the hohlraum target.

The first spherical wave front would heat the target to plasma, and the second beam, aimed at a single point on the plasma ball, would cut to the center of the target, ensuring extra heating to the core and igniting a chain reaction. This scheme, fast ignition, may provide a new promise for laser fusion.

HiPER is still in the planning stage, and engineering problems are being hammered out. One unexpected problem is the production of neodymium-glass disks for the laser amplifiers. An estimated 1,300 such disks will be required, but they are no longer being produced commercially.

The laser fusion projects seem exotic and separated from practical application to power production by decades. There are, however, fusion proposals and experiments that make laser fusion seem like a dull, everyday experience.

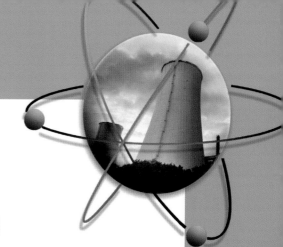

5 Exotic Fusion Reactor Designs

Ever since the disastrous effects of the "War to End All Wars," or World War I, the Treaty of Versailles of 1919 had restricted the building of armaments, ships, and aircraft in Germany. In 1926, the provisions of the treaty were somewhat relaxed, and Germans were once more allowed to do what they did well, build lighter-than-air airships at the Zeppelin Works in Friedrichshafen.

In the years since the war, there had been advancements in airship construction. It had been discovered, to no surprise, that airships filled with hydrogen gas are capable of exploding and quickly burning to a black spot on the ground. Hydrogen gas is an ideal material to use in heavy-lifting balloons and airships, because there is no lighter gas, but it does have a flammability issue. The next best gas, helium, is four times heavier than hydrogen, but it has a grand advantage. It is not flammable. Helium is a chemically inert gas and is incapable of reacting with anything. It is, however, rare. In 1926, the only source of helium was in oil wells in the United States, where it was considered a strategic material and could not be exported.

Two enterprising chemists, Friedrich "Fritz" Paneth (1887–1958) and Kurt Peters (1897–1978), at the University of Berlin proposed a method of making helium from hydrogen using a palladium catalyst. The element palladium is a silvery, semiprecious metal, used in telephone relay contacts, some dentistry, and as a less expensive substitute for platinum in jewelry. It also has an odd affinity for hydrogen. Left alone with hydrogen, it will absorb 900 times its volume in hydrogen gas, until it seems bursting with it. The gas becomes trapped in the palladium, forced in between atoms in the metal's molecular matrix, and subjected to forces many thousands of times greater than atmospheric pressure. Under these conditions, the chemists reasoned, two hydrogen nuclei should combine into one helium nucleus. This would not be a plasma effect, in which the atoms were stripped of electrons and heated up to great collision speeds, but would be a more static process. It would be cold fusion, using a backdoor effect at the low end of the energy spectrum.

The two scientists concocted an experiment in which palladium metal and hydrogen gas were encamped in a sealed glass tube. As predicted, after a waiting period helium gas could be detected in the tube using electrical excitation spectrometry. Putting a few thousand volts of electricity through the tube caused the gas inside to light with two mixed colors, one for hydrogen and one for helium. This exciting finding was quickly published in *Die Naturwissenshaften*.

Unfortunately, the two triumphant chemists had fallen into the helium trap that has tripped many otherwise good experiments. There is a small amount of helium always mixed in with air, as radioactive elements in the soil slowly decay. Alpha particles from radium decay, for example, are actually helium nuclei, and they are a natural background radiation. Helium is so light, there is no buildup, as the atoms gradually float to the top of the atmosphere and are blown off into space by the solar wind. Helium is a small atom, and it can diffuse through almost any material, including the walls of a steel helium tank and especially the walls of a thin, glass tube. The sealed space inside the Paneth and Peters apparatus had become contaminated with helium out of the air.

Paneth and Peters were good scientists, and as such they soon realized their mistake and in the next issue of *Die Naturwissenshaften* they published their short letter of retraction. There was no harm done in thinking that nuclear fusion was possible, and the two chemists went on with their long, productive careers.

This was only the beginning. The Paneth and Peters experiment happened more than once, as the original work was forgotten. Throughout the 20th century, many plans and theories for fusion, sometimes hoping to defeat the tyrannical necessity of plasma conditions, would erupt. Some would crash and burn like a Zeppelin full of hydrogen, and some are still in effect. None have led to fusion power production, but all are emblematic of human ingenuity and a deeply held need for a clean, unlimited supply of power.

MUON-CATALYZED FUSION

The biggest problem in initiating the fusion of two atoms is the repulsive electrostatic force that holds them apart. The first barrier is the electron clouds covering the nuclei. Their negative charges cause incoming atoms to bounce off like nerf balls. Raising hydrogen to plasma temperatures strips off the electrons and eliminates this problem, but it also introduces the problems of plasma containment, which have plagued fusion reactor design for several decades. There is a way around this. Hydrogen has a natural tendency to form a molecule in which two nuclei share electrons, and in this configuration a deuterium nucleus and a tritium nucleus can be close together. In this configuration, the second barrier is encountered. The nucleus of the hydrogen isotope is positively charged, and it prevents nuclei from getting close enough to fuse, even when they share the same molecule.

However, there is another way to build a hydrogen atom. The electron is a very lightweight particle, and it has a very high orbit around the heavy proton. There is another subatomic particle that has the same negative charge as the electron, only it weighs 200 times more than the electron. It is the muon, discovered in 1936 in studies of cosmic ray showers. High-speed protons crashing into the atmosphere from outer space release showers of debris containing muons. They are thought to be part of the glue that holds atomic nuclei together. Substitute muons for the electrons in hydrogen isotopes, and they form extremely compact molecules. The muon flies low around the nucleus, and it is so close its negative charge cancels out the positive charge that normally keeps nuclei apart. A muonic hydrogen molecule is 207 times smaller than a normal hydrogen molecule, and the two nuclei are sufficiently close that the strong nuclear binding force takes over and pulls the nuclei together. They fuse immediately, releasing the energy of fusion and fusion products in a microscopic explosion. It has all the advantages of plasma fusion, without the need to expend a great deal of energy heating the hydrogen to plasma temperature.

The muons are reusable. When a molecule explodes under fusion, the muon used to substitute for an electron flies away as debris, eventually crashes into another normal molecule, and dominates it, pushing out the lesser particle, the electron. The newly equipped molecule collapses under the heavy orbiting particle, and a new fusion occurs.

Muon-catalyzed fusion had been predicted since before 1950, but it was first recorded in a laboratory experiment by Luis Alvarez (1911–88) at the University of California, Berkeley, in 1956. Serious speculations concerning energy production soon followed.

Fusions of deuterium-tritium molecules by muon substitution for electrons could be an inexpensive, pollution-free means of generating power without the severe technical difficulties associated with all plasma-fusion schemes. The muon fusion occurs at room temperature, without special containment issues. A by-product of deuterium-tritium fusion is free neutrons, and these particles could be put to practical use transmuting thorium-232 into fissile uranium-233, thus producing a double-mode energy production. Such a reactor could develop power at both ends of the nuclear mass-spectrum, fusing hydrogen and fissioning uranium. The fission reactor would not even have to be a critical assembly of uranium, but would produce power by fissions triggered by the stream of excess neutrons and not by the usual chain reaction.

There are a few practical problems that will require further research and development. Muons are difficult to generate efficiently. They are released from nuclear matter by simulating a high-energy cosmic ray event in the upper atmosphere using an Earth-bound particle accelerator. These machines are notoriously inefficient, and the energy needed to produce a stream of usable muons is greater than can be extracted from muon-catalyzed fusion. A more economical means of generating muons is needed. It takes about 6 billion electron volts of energy to make one muon using current technology, and one fusion releases 17.6 million electron volts of energy. To break even in energy production, each muon would have to produce about 341 fusions. With that level of efficiency, as much power would be produced as was required to run the reactor.

Muons do not last very long outside the nucleus, with a mean lifetime of about 2.2 microseconds. They decay into two neutrinos and an electron, none of which are useful in a fusion process. In that 2.2 microseconds, it is important to get as many fusions as possible using one muon, reducing the number of energy-expensive muons needed to make power. It is possible to get 100 fusions out of a single muon before it decays.

Along with the limited muon lifetime, there is another phenomenon that keeps the number of fusions per muon low. The notorious alpha-sticking problem has plagued muon-catalyzed fusion research since it was first observed in 1957. The product of a fusion is always at least one new nucleus, different from the original components. A product of the preferred deuterium-tritium fusion mode is a helium nucleus, or alpha particle, sent off into the mixture of molecules waiting to be fused, fusion debris, and free muons. The helium nucleus can latch onto a muon, becoming a super-compact helium atom and denying a muon to the fusion process. Helium atoms, even the compact, muon form, do not form molecules with anything and therefore have nothing with which to fuse. A muon-invaded helium is worthless for energy production, and its muon is rendered unusable. Recent experiments at reducing the alpha-sticking problem indicate that it may be possible to reduce it to the point where as many as 333 fusions could be caused by one muon. This is still short of the 341 fusions per muon needed for a break-even energy process. Work continues to explore the interesting possibility of muon fusion.

THE MIGMA FUSION REACTOR

Bogdan Castle Maglich (1928–   ) was born in Sombor, Yugoslavia. At the age of 12, he and his mother escaped a Nazi concentration camp. He went on to earn a bachelor of science degree from the University of Belgrade in 1951. He achieved a master of science degree from the University of Liverpool, United Kingdom, in 1955, and a Ph.D. in high-energy physics and nuclear engineering from the Massachusetts Institute of Technology in 1959. He joined Luis Alvarez at the University of California, Berkeley, in the discovery of the omega meson. After distinguished positions at the University of Pennsylvania and Rutgers University in New Jersey, he decided to pursue research in the private sector.

In 1972, he completed work on a unique type of hydrogen fusion reactor, named the Migma. To achieve fusion, the problem has always been the coulomb barrier, or the electrostatic force that separates two atoms or nuclei. Maglich determined that the set of nuclei with the least problematic coulomb barrier was a deuterium-tritium pairing. There is only one unit of charge per nucleus, and the addition of three neutrons between the two nuclei adds favorably to the strong nuclear force that causes them to fuse together. An energy of only 100 thousand electron volts is necessary to cause them to fuse and release 17.5 million electron volts. To make these nuclei fuse, he found that he could accelerate two beams of charged particles, one of deuterons and one of tritons, or deuterium and tritium ions, and intersect them. Where they were able to collide, fusion occurred.

This had been tried before, possibly, by Ronald Richter in the much-maligned Huemul Project in Argentina, although it was difficult to reconstruct Richter's experiments because of his lack of documentation. Maglich added an interesting wrinkle. The problem with making break-even fusion using this technique was the very low probability of two ions colliding in the relatively empty space of the collision point. To address this deficiency, he directed his accelerated ion beams into a strong, vertical magnetic field, and this caused the ions to orbit around in a circle. Therefore, if an ion missed hitting another one in the first passage, there were other chances, as it spun around and set itself up for another collision event. The only problem with this stratagem is that the ions that are trying to intersect, the deuterium and the tritium, are both going in the same direction.

With some promising success in 1972, Maglich proceeded to build the Migma II in 1975, the Migma III in 1978, and his final attempt at break-even fusion, the Migma IV, in 1982. These devices were unique in the world of plasma fusion in that they were built without large government grants, and they were relatively small. The accelerator beam-lines were only a few yards long, intersecting in a disk-shaped target chamber about six feet (2 m) in diameter and three feet (1 m) thick. For its small size and expenditure, Migma IV was able to achieve in 1982 a record fusion "triple product," or the product of plasma density, plasma confinement time, and energy produced. Nothing could approach the record until the JET reactor in England got close in 1987. Still, there was no break-even energy production, and private research money dwindled.

A weakness of the Migma design is the fact that the charged particles, or ions, are forced into tight, circular orbits. The transition from a straight-line path into a tight turn is an acceleration, and this causes the particles to emit X-rays, or bremsstrahlung. In radiation X-rays, the particles lose energy, possibly below the fusion point, and this action tends to quench the energy-producing reaction. Attempts by Bogdan Maglich to gain government-sponsored or private funding for further experiments with the migmatron have yet to yield positive results.

COLD FUSION

At the University of Lund, Sweden, in 1927, a Norwegian-born physicist and chemist named John Tanberg (1896–1968) became interested in a paper, "On the Transmutation of Hydrogen to Helium," recently published in the German scientific journal *Naturwissenshaften*. It described a fusing of two hydrogen atoms into one helium atom, accomplished using a palladium catalyst. Tanberg saw more than helium synthesis in this process. He saw the potential for significant power generation, and he thought up a way to improve the performance of the German experiment.

Instead of introducing hydrogen gas to a piece of palladium, Tanberg rigged up an electrolysis cell. Water was kept in a glass beaker, and in the center was dipped an electrode made of palladium. The palladium became the cathode, or the negatively charged electrode, in an electrical circuit. The anode, or positive electrode, was a copper wire dangling in the water. A small current of electricity from a battery was applied across the electrodes, and hydrogen gas bubbles developed on the cathode. Oxygen gas, separated from the hydrogen in the water, escaped off the anode. The hydrogen was soaked up by the palladium, forced into the small spaces between palladium atoms making up the cathode metal, and Tanberg detected fusion taking place. The temperature of the water in his electrolysis cell rose as the palladium soaked up the hydrogen.

Elated by his finding, Tanberg immediately filed for a Swedish patent. The application was returned with the message "read but not understood." Tanberg's patent was rejected, as the examiner found the description of his apparatus too sketchy. This was disappointing but not discouraging. In 1932, deuterium, a heavy isotope of hydrogen, was discovered and more had been learned concerning the structure of atoms. Tanberg obtained a sample of heavy water from the Norsk Hydro fertilizer plant in his native Norway and tried the electrolysis experiment again, with apparently better results. Still, Tanberg was unable to generate interest in his palladium electrolysis cell as a power source. He eventually abandoned the work and became science director of a company making vacuum cleaners, Electrolux, in Stockholm.

In 1989, history repeated itself, this time with much more attention but with a similar outcome. Steven E. Jones, a physics professor at Brigham Young University in Salt Lake City, Utah, had been studying the anomalous concentration of helium-3 coming out of volcanoes, and he had taken samples of the gas emerging from the ground in Hawaii. Helium-4 in a volcano vent or an oil well was easy to explain. As certain radioactive isotopes in the Earth decay, some of the radiation is in the form of alpha particles. An alpha particle is a helium-4 nucleus. All it has to do is find two loose electrons and it becomes a bona fide helium atom. In any break in the Earth's crust, such as an erupting volcano, the helium-4 that has collected for millions of years will escape easily to the surface. Helium-3 is another matter. No helium-3 was thought to be included in the material from which the Earth was made, and the only way to make new helium-3 is with fusion.

Jones questioned how fusion could occur inside the Earth. The Earth is, at the core, hot enough to melt rock and under great pressure, but this is nowhere near the conditions necessary for plasma fusion of hydrogen isotopes. It is probably no hotter than 9,800°F (5,700°C) at the center of the Earth, and the helium-3 was coming from just under the surface. A possible fusion that would make helium-3 could be deuterium-deuterium. Deuterium exists in nature, but it is not common. About one hydrogen atom out of every 6,400 is deuterium. Somehow, two deuterium atoms would have to find each other and fuse together.

Jones came up with a possible explanation for the presence of helium-3. There are compounds known as hydrides, by which hydrogen atoms are attracted to certain metals and tend to lodge in between the metal atoms, as if they were pressed in at tremendous pressure. Palladium is an ideal hydriding material, but next down on the limited list of hydriding metals is nickel. The center regions of the Earth are made of a nickel-iron mixture. Perhaps deuterium atoms were being forced to fuse by the hydriding process in nickel. Jones prepared a theoretical paper by March 1989.

Also in Salt Lake City is the University of Utah, where Stanley Pons (1943–   ) was chairman of the chemistry department. He and a chemistry professor from Salisbury, England, Martin Fleischmann (1927–   ), had applied for funding from the U.S. Department of Energy to investigate the fusion of deuterium in a hydriding metal, palladium. They had funded their own research with $100,000 out of pocket and were looking for more money for further study of what they considered an intriguing effect. Being electrochemists, they had tried electrolysis of heavy water as a first step in their research, with interesting results. Pons, Fleischmann, and Jones had never heard of Tanberg's identical work 50 years before.

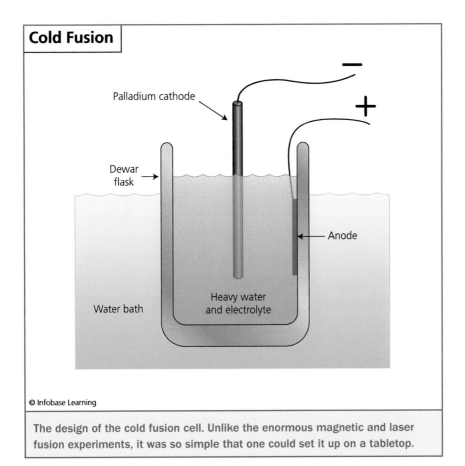

Cold Fusion

Palladium cathode

Dewar flask

Anode

Heavy water and electrolyte

Water bath

© Infobase Learning

The design of the cold fusion cell. Unlike the enormous magnetic and laser fusion experiments, it was so simple that one could set it up on a tabletop.

One of the peer reviewers for their grant proposal was Steven Jones, and he soon met with Pons and Fleischmann to compare notes on their amazingly similar theories about fusion in hydrides. The researchers from both universities were preparing papers for publication on this topic, and in a meeting on March 6, 1989, they agreed to publish simultaneously, on March 24, out of professional courtesy. Pons and Fleischmann, under pressure from higher levels of management in the university, jumped the gun and sent their paper to the *Journal of Electroanalytical Chemistry* on March 11. As the management knew, funding from external sources goes to the swiftest. Breaking all tradition and outside scientific protocol, they held a press conference on March 23 to announce their experiment and findings to the entire world. In front of the news cameras, Pons and Fleischmann said that they had

A cold fusion apparatus being assembled. The key to cold fusion was the palladium cathode, shown in the center with a wire attached. *(Philippe Plailly/ Photo Researchers, Inc.)*

induced deuterium fusion in a glass bottle on a table-top, making "tremendous heat" and "detectable neutrons." Jones, who had held his paper as agreed, was upset.

News of this discovery was something of a shock to the scientific community. Fusion power research was ongoing worldwide, with billions of invested dollars, and this announcement out of Utah made it appear that all the fusion scientists were looking in the wrong place for power production. Two electrochemists in Utah had self-funded a fusion experiment that made more power than it used, something that no plasma fusion device had ever done. The pair from Salt Lake City had found a loophole in nature, on the extreme other end of the energy spectrum where others had not looked, making "cold fusion" instead of hot, plasma fusion. There was an immediate rush to duplicate the Pons and Fleischmann experiment to confirm their outlandish findings. Hundreds of laboratories worldwide obtained palladium for electrodes and began assembling electrolytic apparatus.

It seemed odd that in news reports Pons and Fleischmann were shown standing over their fusion cell, leaning over to peer into it while it gener-

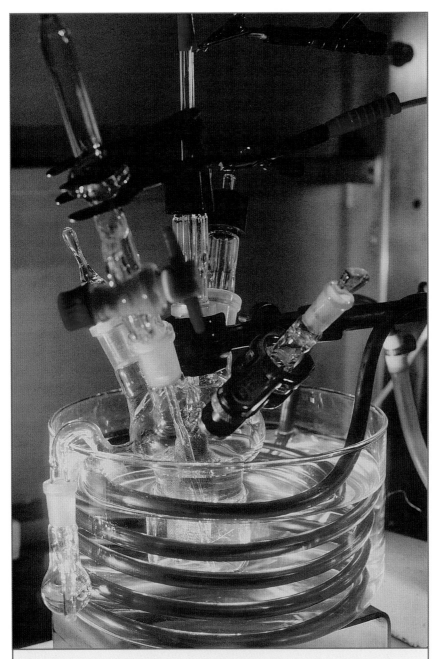

An elaborate French apparatus being used to investigate the results of the
Pons and Fleischmann experiment. The platinum-titanium cell is cooled
by a refrigerant flowing through the copper pipe. *(Phillipe Plailly/Photo
Researchers, Inc.)*

ated a "tremendous amount of heat." Under the conditions of deuterium-deuterium fusion, a tremendous amount of heat production would result in a more than tremendous production of neutrons, and it would be dangerous to stand so close to a neutron-generating process. It would have made more sense to claim a tremendous neutron production and detectable heat.

By early April 1989, positive and negative results of replication efforts were reported in the news. Most laboratories could find no evidence of fusion in electrolysis cells, but a team at Georgia Institute of Tech-

THE FARNSWORTH-HIRSCH FUSOR

Philo Taylor Farnsworth (1906–71) is best known as the inventor of electronic television. Rudimentary television receivers and transmitters had been built using motor-driven disks to break a picture into lines or pixels for transmission, but Farnsworth conceptualized a fully electronic system of image dissection as a high school student in Rigby, Idaho. Farnsworth had a natural gift for electronics theory and design, and by 1927 he had set up a laboratory in San Francisco, California, and demonstrated his first electronic television imaging device. It was a specially built vacuum tube of unique and ingenious design.

By 1930, Farnsworth was investigating a number of radical vacuum tube designs for use in television, including an amplifier tube he named the multipactor. This tube could amplify an incoming signal by a factor of millions as it moved from one electrode in the vacuum to another. Farnsworth found that he could stop the electrons in mid-flight using a magnetic field alternating polarity at a high frequency. A cloud of electrical charge would accumulate in the middle of the tube. It was an interesting effect, and it did amplify signals, but it also eroded the electrodes quickly. Farnsworth's multipactor effect became known as something to avoid when designing a vacuum tube.

After decades of patent fights and settlements with RCA over the ownership of television, Farnsworth Television Labs was bought by the ITT Corporation in 1949. Farnsworth was always finding new ways to use a vacuum tube, and by 1960 he became interested in fusion power developments. A problem that was plaguing all fusion experiments was the effort to contain the plasma and keep it from hitting the walls of a fusion vessel. Farnsworth recalled his experiences with the multipactor,

nology in Atlanta, Georgia, claimed to have recorded slight neutron activity plus tritium production, both measurable attributes of deuterium-deuterium fusion. No heat production was notable. A separate team at Texas A & M University in College Station, Texas, reported heat production measured by calorimetry. Within days the Georgia Tech team retracted their finding, having found problems in their neutron detector. The A & M team became quiet and issued no further claims. Negative results were reported widely. In May 1989, cold fusion was proclaimed dead by a panel of nine scientists from the American Physical

thinking that he had a way to isolate an electrical charge in the middle of a vacuum tube, and it should work as well with a blob of ionized hydrogen isotopes. Fuel could be injected from a port in the side of his multipactor, and it would be unable to escape. He built several examples, and named the design the *fusor.*

It worked. Tritium and deuterium ions would be trapped in a cloud formation at the center of the device, spinning around in a sort of mini-tornado, occasionally bumping into each other and fusing. The proof was a flux of neutrons streaming out of the device. There was a problem. The design did not seem to scale up to a fusor that would actually generate usable power. Although it was definitely causing hydrogen isotopes to fuse, in the tiny laboratory experiment it required much more power to operate the fusor than could be extracted from it. An entire fusor would fit on a kitchen table.

Another scientist, Robert L. Hirsch, arrived at the lab with some fresh ideas. He redesigned the fusor, using nested, spherical wire-cage electrodes in a spherical vacuum chamber. The new, improved design became known as the Farnsworth-Hirsch fusor. Soon the device was exhibiting neutron production rates up to a billion per second. More work would coax a trillion fusions per second from the fusor.

In the middle of 1967, just when the fusor was beginning to be impressive, ITT ended the funding for the experiment. The company had gotten out of the telephone and electronics business back in 1961 and had moved to buy insurance companies, Sheraton Hotels, Wonderbread, and Avis Rent-a-Car. Building a future power source was not in the expansion plans. Farnsworth moved to Brigham Young University in Salt Lake City, Utah, and tried to resume his fusion research. All his money went into the venture, his savings were soon used up, and he died of pneumonia in March 1971. As a source of power, the fusor project died with him, but fusors are built today as neutron generators by NSD-Fusion in Delmhorst, Germany.

Society, and negative papers followed in several leading scientific journals. In November, a panel at the U.S. Department of Energy voted not to fund cold fusion research.

In 1992, Stanley Pons and Martin Fleischmann moved to France to continue their studies of cold fusion with funding from Toyota of Japan. The laboratory closed in 1998, showing no positive results after spending $18 million. Pons became a French citizen, and Fleischmann returned to Salisbury, England. Steven Jones became famous for his strongly held theory that the World Trade Center in New York was felled by controlled demolition during the morning of September 11, 2001, and for this he was relieved of teaching duties and placed on paid leave from Brigham Young University. The U.S. Patent Office refuses to grant a patent to an invention that mentions the phrase "cold fusion" or references work by Pons and Fleischmann. The International Conference on Cold Fusion has held yearly meetings since 1990.

POLYWELL FUSION

Robert W. Bussard (1928–2007) worked for more than 60 years in nuclear energy research, starting in 1955 at the Los Alamos National Laboratories in New Mexico. He joined the nuclear propulsion division and worked on Project Rover to develop a nuclear rocket engine. In 1960, he came up with his own deep-space propulsion design, employing a ramjet that would collect space-borne hydrogen ions in front of the vehicle and use hydrogen fusion to propel forward.

Bussard picked up the idea of improved fusion plasma confinement in 1983 from work that had been performed in the late 1960s, including an intriguing fusion system designed by Philo Farnsworth. Farnsworth's fusor was designed to confine hydrogen isotope ions in a cloud in the center of a small space and encourage them to collide with each other and fuse. The Farnsworth fusor had been improved into the Elmore-Tuck-Watson fusor, but even this advanced design could not achieve energy production by fusion. Bussard saw the problem. The fusor confined the ions electrostatically, and the latest design required electrodes to establish the electrostatic field. The presence of electrodes was draining energy out of the plasma as particles hit the electrode grid before meaningful fusion could occur. Get rid of the wire-basket electrodes, and the fusor would generate usable power.

Instead of the metal wire electrodes, Bussard proposed virtual electrodes, which actually was a throwback to Farnsworth's original fusor multipactor vacuum tube design from the early 1930s. This was an important modification of the current fusor implementation. In Bussard's first fusion experiment, named the WB-1, or "wiffleball-one," polywell, his confined plasma cloud was defined by a polyhedron. A polyhedron is a three-dimensional geometric solid with flat faces and straight edges. The simplest polyhedron is a three-sided pyramid made of triangles. The first polywell was made using the next highest polyhedral shape, the cube with six faces. Each face was composed of a donut-shaped coil of wire, encased in a metal shell, with each one acting as a magnetic mirror to reflect charged plasma back into the center of the polyhedron. This polyhedral collection of magnet coils, called the MaGrid, was inside a sealed metal chamber, standing off the bottom on four ceramic insulators.

In operation, an electrical charge was induced between the metal chamber and the MaGrid, with the coils charged positive and the chamber charged negative at about 12,500 volts. Electricity was applied to the magnets, creating a magnetic field focused on the center of the polyhedron, air was vacuumed out of chamber, and deuterium gas was introduced through a side port. Incrementally improving the design in WB-2 through WB-6, Bussard claimed to have achieved fusion rates 100,000 times greater than Farnsworth ever achieved with his fusor under similar conditions. By 2005, the polywell was producing a billion fusions per second. Bussard calculated that the machine would produce net power if it were scaled up to be 10 times larger and if superconducting magnets were used to produce the magnetic fields, just as his project funding ran dry. He had plans for a WB-7 and a WB-8. The MaGrid in WB-8 was to be built as a higher-order polyhedron having 12 sides, giving an improved smoothness to the shape of the magnetic field derived from the coils.

Initial funding was provided by the Department of Defense starting in 1987. In 1992, the U.S. Navy provided low-level funding, and in 2005 the Office of Naval Research gave $900,000 to complete the WB-6. The fusion rate achieved in this device was impressive, but only in sub-millisecond pulses. Bussard predicted continuous, high-level fusion in a larger model which would cost no more than $200 million dollars to build. At this point, the insulation in one of the hand-wound coils on the MaGrid burned through and the machine destroyed itself. Government funding ceased.

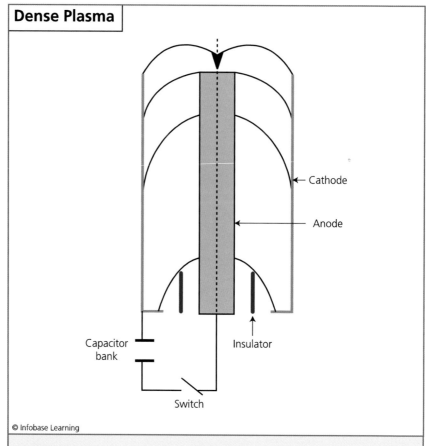

Dense Plasma

Cathode

Anode

Capacitor bank

Insulator

Switch

© Infobase Learning

A schematic diagram of the dense plasma device. It is very simple, consisting of a metal tube with a metal rod down the center. The black lines show how the shock wave develops as it moves through the tube, from bottom to top and out the end.

In 2007, Bussard established a nonprofit organization to gather donations and continue work on the polywell, the EMC2 Fusion Development Corporation. Later that year, Bussard passed away. The WB-7 was completed in 2008 by EMC2; a WB-8.1 is expected to be completed in 2011; and there are plans for a commercial power station by 2020.

THE DENSE PLASMA FOCUS

A machine that produces a short-lived burst of hot, concentrated plasma was developed independently in the United States and the Soviet Union in the early 1960s. This plasma gun consists of two metal electrodes, cylindrical, with the smaller-diameter cylinder nested inside the larger-diameter cylinder. The hollow inner cylinder and the space between the small and large cylinder are filled with a gas. A bank of electrical capacitors is charged to a high voltage, in the 100,000-volt range. Switch the charged capacitors to discharge across the electrodes, and the gas instantly breaks down into a plasma. The electrodes are shorted together by the superconducting property of the plasma, and a magnetic Z-pinch effect develops, similar to that used in the ZETA fusion reactor experiments in the United Kingdom.

Under the influence of the intense magnetic field derived across the electrodes, the plasma is accelerated along the main axis of the cylindrical electrodes, shooting out the end of the device, focused down to a slender beam of glowing material traveling at many times the speed of sound. The pulsed event is short and loud, lasting only a few microseconds.

The plasma gun can be small enough to fit on a tabletop, yet it can be scaled up to a machine that fills an entire building. The physical size and the power output can be increased, but the pulse-length remains slight. Unlike other Z-pinch experiments, the dense plasma focus device makes no attempt to control the extreme instability of magnetically confined plasma. The goal is not to sustain the reaction, but to make the pulse as energetic as possible. Instead of eliminating the instability, it is exploited, forming an imploding shock wave and a dense column of short-lived plasma.

A hot, dense plasma is an ideal environment for nuclear fusion. Fill the plasma gun with deuterium gas, and a sudden, intense burst of neutrons emerges from the device as the electrical power is applied to the electrodes. This phenomenon indicates that deuterium-deuterium fusion is occurring in the plasma jet. About half of all deuterium-deuterium fusions results in a helium-3 nucleus, plus one stray neutron, whereas the other half results in one tritium nucleus plus one free proton. Of these possible fusion products, neutrons are the easiest to detect, and their appearance confirms that there is definitely fusion action in the plasma as it projects out the front of the device.

A practical use for the dense plasma focus device is to test incoming shipments from overseas for smuggled nuclear weapons or fissile weapon components. All fissile isotopes used in weapon construction are at least slightly radioactive, but the alpha, beta, and gamma radiations are easily shielded against detection by radiation detection and counting equipment. Neutrons, however, are not affected by anything used as a radiation shield. They are uncharged particles and are free to move through most solid material. Release a pulse of neutrons from a dense plasma focus, and it will penetrate any shipping container plus shield material, bathing a hidden atomic bomb with neutrons. Any fissile material in a bomb, be it uranium or plutonium, will react to the incoming neutrons by fissioning, releasing more neutrons. These neutrons are free to penetrate the shipping container and all shielding material going the other way, from the bomb to an outside neutron counter. A proposed bomb detector uses a dense plasma focus to reach a hidden nuclear device with neutrons. Any neutrons detected in excess of the plasma-pulse are considered indications of a hidden bomb.

The concept of a fusion power plant based on dense plasma focus has been studied since the 1960s. Such a fusion reactor will have to run in pulse-mode, generating power in repeating, short bursts, with a building full of electrical capacitors being charged and then discharged into the machine. The plasma developed in this type of machine is sufficiently dense to possibly allow the rarely proposed hydrogen-boron fusion. This particular fusion results in no radioactive isotopes and no neutrons, making it a near ideal power source producing no radioactivity.

Although it may be possible to generate practical power using a modified plasma gun scheme, this technology suffers from the same problem that faces many plasma-fusion designs. There must be metal electrodes in the gun to establish the ionization by electricity and the resulting magnetic field. Anything solid in the way of the plasma, even essential electrodes, will stop flying particles while producing no usable energy. A plasma particle can slam into an electrode and generate a high-energy X-ray, and this drains energy from the plasma. Research is underway to overcome this limitation, and the latest dense plasma focus fusion device built is the FoFu-1 at Lawrenceville Plasma Physics in Middlesex, New Jersey. There is dense plasma focus research ongoing in Poland, Malaysia, Argentina, and at Kansas State University, Kansas.

BUBBLE FUSION

At the University of Cologne in Germany in 1934, there was a great deal of research into sonar, and as a result there was no shortage of ultrasound transducers. These devices, usually consisting of a precision-ground crystal with two wires attached, were used to create sound waves of a frequency so high no human could hear them. They would work underwater, and development was underway to use ultrasound to pinpoint the positions of submerged submarines.

A pair of scientists involved in the sonar project wanted to speed up the process of developing photographic film, and they thought to sink an ultrasonic transducer in a tank of photo-developer to give it extreme agitation. To their surprise, the film came out of the developer with tiny dots of exposure, as if a thousand points of light had been turned on during the developing process. Investigation revealed that ultrasonic waves in a liquid medium have a curious effect on bubbles. These miniscule voids are slammed shut by the standing sound wave in the liquid, and as they collapse the bubbles emit a short pulse of light. This phenomenon is referred to as multi-bubble sonoluminescence.

The reason why bubbles forced to collapse in a liquid emit light remains open to speculation. There are multiple theories. It is known from measurements that extremely high-temperature plasma develops in the bubbles as they implode, as hot as 100,000 degrees. The bubble temperature could even reach 10 million degrees, as hot as the center of the Sun, and this possibility came to the attention of nuclear fusion specialists. The fusion of deuterium in sonically induced bubble-collapse became formally known as acoustic inertial confinement fusion, or simply bubble fusion.

Although a patent for bubble fusion was filed as early as 1978, the phenomenon reached international notoriety in 2002 when Rusi P. Taleyarkhan and associates at the Oak Ridge National Laboratory in Tennessee announced successful bubble fusion in a glass container full of acetone. In the acetone, which is a hydrocarbon solvent, the deuterium isotope of hydrogen had been substituted for the ordinary hydrogen, making possible a deuterium-deuterium fusion under extreme conditions. For bubble fusion to occur, the bubbles must be extremely small, and simple air bubbles will not work. To make appropriately sized bubbles in the deuterated acetone, Taleyarkhan bombarded the glass container with a stream of neutrons, acting as seeds for the tiny voids in the fluid. An ultrasonic transducer around the middle of the container provided a standing sound wave to crush the bubbles, a microphone picked up the clicking sounds of bubbles collapsing, and a neutron detector was used to measure neutrons produced by the fusion. The expected products of a deuterium-deuterium fusion are neutrons and tritium atoms, produced in approximately equal numbers. By

A photo, taken on May 31, 2002, shows Rusi Taleyarkhan experimenting with bubble fusion at the Oak Ridge National Laboratory, Tennessee *(DOE Photo)*

noting the coincidence of a neutron activating the neutron detector and the click of a bubble implosion, Taleyarkhan could confirm each fusion incident. His team reported definite evidence of fusion occurring in the deuterated acetone under ultrasonic sound bombardment. The effect was not an overwhelming production of energy, but it was sufficient to indicate that further investigation was appropriate.

Taleyarkhan prepared another experiment identical to the first but using ordinary acetone without the deuterium as the bubble fluid. From this control experiment he recorded no nuclear activity. This further indicated that deuterium-deuterium fusions were occurring in the deuterated acetone.

Later in 2002 a separate team of scientists at Oak Ridge repeated the experiment independently, using improved neutron-counting equipment. They reported finding no noticeable neutron production in addition to the confusing neutron production by the source used to seed bubbles in the fluid. Taleyarkhan rebutted this finding, pointing out differences in this experimental setup and his original arrangement of equipment. In 2004, Taleyarkhan staged another bubble fusion experiment, this time reporting improved fusion indications.

In 2005, the British documentary series *Horizon* commissioned a new bubble fusion experiment by leading sonoluminescence researchers, using a replica of Taleyarkhan's equipment with much more sophisticated neutron detection. They could find no hint of neutron production, and therefore they reported no fusion. Later in 2005, two of Taleyarkhan's students at Purdue University published the results of an experiment confirming his results. This experiment, while bolstering bubble fusion, unfortunately led to research misconduct charges against Taleyarkhan. He had deep involvement in the experiment, yet he was not named in the author list, indicating less independence of the findings than was implied. Physical phenomena that are so fleeting and sensitive that they reveal themselves only in the presence of one individual generally have a weak connection to reality.

Atop these contradictory findings, in 2005, three Russian scientists reported a neutron yield in deuterated glycerine under shock compression by ultrasonic sound. In 2006, a separate group of researchers at Purdue University reported neutron emission in deuterated solvents mixed with uranyl nitrate. This experiment used no external neutron source to make bubbles, with alpha particle emissions from the uranium compound providing void-seeding. Confirmations, anti-confirmations, allegations, and

Bubble Fusion

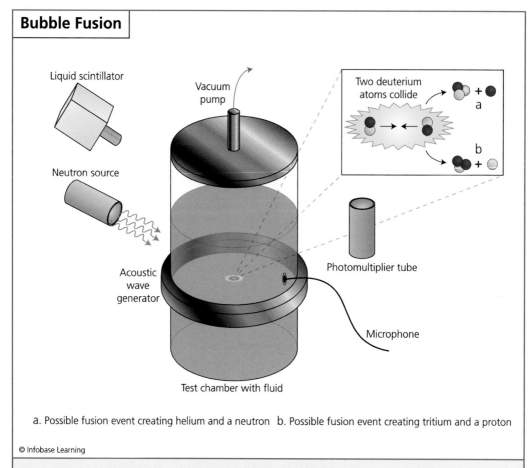

a. Possible fusion event creating helium and a neutron   b. Possible fusion event creating tritium and a proton

© Infobase Learning

The bubble fusion experiment is as simple as a cold fusion setup and equally controversial. Externally produced neutrons form microscopic bubbles in deuterated acetone, and they are forcibly collapsed by ultrasonic acoustic waves. The bubbles supposedly collapse with sufficient force to cause deuterium-deuterium fusion.

investigations flew back and forth for six years after Taleyarkhan's initial paper. On August 27, 2008, Rusi Taleyarkhan was stripped of his Arden L. Bement Jr. professorship and forbidden to be a thesis adviser for graduate students at Purdue University.

On December 5, 2007, the Kimberly-Clark Corporation of Irvine, Texas, filed for a patent on a device that appears to be a bubble fusion reactor. The patent describes a thermonuclear fusion reactor using an ultrasound generator aimed at heavy water, or deuterium oxide. Kimberly-Clark manufactures consumer paper products, including Kleenex tissues and Cottonelle toilet paper.

THE PACER PROJECT

This last topic in the list of exotic fusion reactors is unique in that it is guaranteed to produce more energy than is expended in the process, using thermonuclear fusion as the power source. Issues of capital cost and physical security have prevented its implementation to this point. While the *PACER project*, conducted at Los Alamos National Laboratories around 1975, could generate significant power, the per-kilowatt expense would make it impractical compared to conventional costs of electricity.

The proposed system was to consist of an underground, cylindrical chamber, 100 feet (30 m) in diameter and 300 feet (100 m) tall, made of a steel alloy 13 feet (4m) thick. It was to be built in an excavated chamber, cut into the bedrock of Nevada and secured with hundreds of 45-foot- (15m-) long bolts driven into the surrounding rock. The cylinder would then be partially filled with molten fluoride salt to a depth of 100 feet (30m). A waterfall of molten salt would be circulated, falling in from the top of the chamber to the pool below and circulated by powerful pumps in an adjacent rock formation. The total volume of salt pumped around the circuit would weigh 2,200 tons (2,000 tonnes). The salt would be routed through a steam generator, converting the heat in the molten lithium fluoride into steam to run a turbo-generator. The end product would be electrical power.

Energy to melt the salt and ultimately run the turbo-generator was to be provided by dropping a one-kiloton (4×10^{12} j) thermonuclear bomb into the chamber and setting it off. A fresh bomb would be introduced and detonated every 45 minutes. The PACER system design was based on data collected during the GNOME test in Project Plowshare in 1961, in which a small atomic bomb was detonated in a salt dome under Carlsbad, New Mexico.

For such a system to operate, a continuous supply of small hydrogen bombs would have to be delivered to the power plant, making the economics of this scheme questionable. Although large-volume mass production would surely bring down the cost of manufacture, hydrogen bombs have never been inexpensive. The model MK28RI internal-parachute fall-retarded thermonuclear gravity bombs, for example, retired from the U.S.

arsenal in 1991, cost about $1 million apiece in 1964. The design of a special nuclear explosive device having a fusion component with an energy release of only one or two kilotons of TNT was challenging. Due to security concerns, plans were to build the bomb factory on the same restricted area containing the blast chamber and power-generating equipment.

As is the case of many fusion power production schemes, PACER is tantalizing, but it remains outside the scope of immediate application. Plans for the future of fusion power have been concentrated into two distinct areas of development, and one or both designs could be operable sometime in the future. These systems, currently in construction, are discussed in the next chapter.

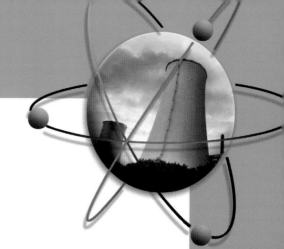

6 The Future of Fusion Power

The Large Hadron Collider (LCH) is the world's largest synchrotron particle accelerator. It is also the most expensive science experiment in history, with a budget of $9 billion. The facility is managed by CERN, the European Organization for Nuclear Research, and its circular tunnel, 17 miles (27 km) in circumference, straddles the border between Switzerland and France. With its construction completed in 2009, the LHC is capable of running two speeding protons into each other, head on, at a combined energy of 7 trillion electron volts. The ultimate purpose of this experiment is to recreate the conditions that existed at the beginning of the universe and confirm scientific predictions.

As is the case of many particle accelerator construction projects, by the time the LHC was ready for low-power testing in 2008 scientists realized that they would need a bigger one. The LHC, for all its magnificent size, cost, and effort to construct, will not quite take the scientists where they want to go, and a design was started for the next collider. Where the construction of the LHC required the collective science budgets of almost the entire continent of Europe, the next collider, the International Linear Collider (ILC) will require the resource contributions of the entire world. The cost of this proposed project is estimated to be as high as $25 billion.

This phenomenon, in which a bigger unit will be required to meet an improved estimate of scientific goals, seems also true of fusion power research. There have been hundreds of tokamak reactors built, but the experimental finding has always been that a more powerful reactor will be necessary to achieve the ignition point, the condition at which a self-sustaining fusion makes usable power. The same is true of the laser-equipped inertial confinement machines. The next one built will be larger in order to achieve ignition. From these observations, a projection into the future of fusion power indicates that larger, more powerful experimental machines will be built. As the expense of a single experiment grows without an increase in available funds, the areas of research will have to focus down to the two having the most previous development, the magnetic confinement torus and the pulsed laser device. Exotic forms or anything off these tracks will lack the support necessary for practical development.

This behavior of incremental equipment resizing is not unique to fusion power development. Although it would appear much simpler and more straightforward than this in written history, the development of fission power during the early 1940s enjoyed a similar stair-step approach to eventual power production. The quest for a self-sustaining nuclear fission reactor started in 1941 with small assemblies of graphite bricks and uranium fuel in laboratories at Columbia University. With each experiment, the scientists learned more about what would be necessary to achieve ignition, expressed as a "k-effective" greater than or equal to one. As more was observed and measured, the machine had to be made bigger. The program moved to the University of Chicago early in 1942. Larger and larger machines were built, but with each experiment it became clear that an even bigger pile of graphite would be necessary. Finally, in December 1942, the fission team reached ignition, and the uranium-fueled reaction achieved the break-even point. To make usable power, it was later found that a larger pile of bricks and fuel was necessary.

A similar sequence is occurring today with fusion power. The difference is that, under wartime conditions, it took three years to develop fission power. So far, it has taken 60 years to try to do the same thing with fusion.

THE ITER (INTERNATIONAL TOKAMAK EXPERIMENTAL REACTOR)

The *ITER* is definitely part of the future of fusion power. The acronym ITER was created at the project's beginning in 1985; the letters stood for International Thermonuclear Experimental Reactor. This proved to be a poor choice of wording, as the general public connected the term *thermo-nuclear* with hydrogen bombs. The name of the project was changed to Iter, which in Latin means "road." While better, it did not express the scope of the project, and it is presently known as ITER, the International Tokamak Experimental Reactor.

The ITER project began as a collaboration among the Soviet Union, the European Union, the United States, and Japan to collectively fund a grand fusion energy experiment. The objective was to combine the lessons learned from all previous magnetic confinement experiments and build a tokamak that would not only reach the ignition point and produce power, but would give a sustained production interval instead of a single pulse. The successful design features of the best tokamaks in the world, including the EAST in China, ASDEX Upgrade in Germany, the JET in the United Kingdom, and Tore Supra in France, were studied and incorporated into the preliminary ITER design.

The proposed features of this reactor are impressive. It will produce 10 times more thermal energy than is supplied to the torus from external sources and will maintain this condition for up to eight minutes. There is no actual power production, and the heat that is generated will be drained off the fusion chamber to prevent melting and will be discarded into a cooling tower. Still, if successful it will prove that usable energy can be generated by fusion, and the conversion of the heat to electricity can be accomplished by conventional means. The ignition-point temperature of the plasma will be exceeded, beyond which the fusion will sustain itself with no heating applied by external sources. Tritium gas, which is necessary for an advantageous deuterium-tritium fusion, does not occur in nature. It will be demonstrated that a fusion reactor can produce its own tritium by neutron capture in lithium-6 in a blanket surrounding the fusion chamber. The neutrons are a by-product of the deuterium-tritium fusion. Shielding from neutrons not used in the tritium manufacture will be demonstrated. Finally, methods of remotely handling maintenance of the reactor, which will be made radioactive by the neutron production, will be developed, using robotics. Results of the successful meeting of these objectives will then be used to build the world's first fusion power station.

This is a hefty set of goals. The best performance of a tokamak to this point has been the JET, which was able to produce exactly one sub-second burst of 16 million watts of power. ITER is designed to make 500 million

INDIA'S SST-1 TOKAMAK REACTOR

On the banks of the Sabarmati River in the Bhat village of the Gandhinagar district of India is the Institute for Plasma Research, operating under the Department of Atomic Energy. In 1995, after successful experiments with the ADITYA and SINP tokamaks, a decision was made to build a steady-state superconducting tokamak named SST-1. The objective of this bold project is to establish a fusing plasma that is steady for 1,000 seconds. If successful, it will rival the ITER and every other experimental magnetic confinement machine as a model for fusion power generation.

It is not only designed to have a steady-state plasma, it will have a controlled power exhaust, which is the beginning of a system to convert the generated heat to electricity. Auxiliary plasma heating will be provided by neutral particle beams and microwave injection. Cryogenic systems are being built to handle the liquid helium for the superconducting magnets and the liquid nitrogen to cool the double-hulled fusion chamber.

Compared to other high-performance tokamak projects, the SST-1 is rather small. Its fusion donut is only seven feet (2.2 m) in diameter, with a 1.3-foot (0.4 m) hole in the middle. The entire machine, including the magnet coils around the torus, is encased in a welded metal cylinder, filled with liquid helium. The auxiliary plasma heating machinery is located off to the side, feeding into the torus through pipes. The design team hopes to maintain steady-state plasma through a computer-controlled feedback mechanism for the "double-null" plasma divertor. By monitoring and auto-

watts of power for up to 1,000 seconds. To accomplish this, it must be many times the size of the largest tokamaks presently in existence, with a fusion chamber volume of about 30,000 cubic feet (840 m³). As a comparison, the fusion chamber in the JET reactor is only 2,800 cubic feet (80 m³). All the air will be removed from the donut-shaped metal chamber and replaced with 0.018 ounces (0.5 g) of a deuterium-tritium gas mixture.

The central magnet for the tokamak, supplying the bulk magnetic field, will use superconducting niobium-tin, carrying 46,000 amperes of electrical current, producing a field of 135,000 gauss (13.5 tesla). The 19 toroidal field coils, evenly spaced around the donut, will require 41 billion joules of energy to reach the required magnetism. They will also be made matically adjusting the divertor, which scrubs the fusion products out of the plasma, they may be able to avoid disruptions that make the plasma hit the chamber wall.

First operation of the SST-1 was attempted in 2006, but multiple problems showed up. Leaks in the cryogenic helium and nitrogen lines caused the coolant for the magnets to boil away, and superconductivity could not be achieved. Heat loads on the superconducting magnets due to their proximity to the fusion chamber were larger than expected, and thermal shields had to be installed. Leaks opened up in the joints of the magnets, and no amount of epoxy glue would seal them. Insulators leaked where wires entered the cryostat.

The problems were bothersome, but there was nothing that could not be fixed. It was decided to completely dismantle the SST-1 and refurbish it using the lessons learned from the unfavorable startup. This time, all subsystems were to be tested under cold conditions before they were installed in the machine, and there would be some redesigns of key components. After complete disassembly, leaks were found in the joints holding together the fusion chamber, preventing a full vacuum before the hydrogen gas was injected. A rebuild and testing of the chamber was begun in November 2009. The refurbishment is based on meticulous testing, first at the component level, then the subsystem level as components are assembled, and finally at the system level as subsystems are installed, one at a time.

The SST-1 may be ready for operation as early as January 2012. Testing will begin under low power, hoping for a first plasma duration of about 0.3 seconds. The goal of 1,000 seconds and a triumph in the world of plasma physics will be approached with due caution.

of superconducting niobium-tin and will use an unprecedented 80,000 amperes of current.

In 1993, the Soviet Union collapsed and was soon replaced by the Russian Federation as a partner in ITER. Initial cost estimates were around $8 billion. Anticipating unattractive cost increases, the United States dropped out of the project in 1999. Canada dropped in, stating that they would prefer to have the reactor built in Clarington, Canada. Japan argued strongly that it should be built in Rokkasho, Aomori, Japan, and Spain offered a site at Vandellós. France had always assumed that it would be built in Cadarache in Provence-Alpes-Côte-d'Azur. Canada, finding that it had no hope of hosting the experiment, dropped out of the consortium in

2003. The United States dropped back into the project right before Canada dropped out. Japan and the European Union came to an understanding in 2005, and it was agreed that the reactor would be built in France, with Japan getting a supercomputer facility and a small materials-test tokamak in the deal. Japan was also promised that 20 percent of the research staff on-site in France plus the head of the administration would be Japanese. South Korea, India, and the People's Republic of China have signed into the consortium, making it a worldwide effort to create fusion power. Kazakhstan has offered to join the program. The project will employ 600 scientists, engineers, and technicians for 20 years.

In 2006, cost projections for the ITER project had grown to $13 billion. This estimate included the operations costs for 20 years plus the decommissioning and teardown. The European Union will assume 45 percent of the total budget and other participants will each contribute 9 percent. Construction of the special components has been divided up among certain participants. Spain, for example, was given the contract to build 10 of the 19 torroidal field magnets, or "winding packs," at a cost of $200 million. Each winding pack is an enormous part, weighing 100 tons (90 tonnes). A Japanese company won the contract to build the remaining nine units.

The ITER project is not without criticism, particularly in the United States. Proponents of non-tokamak fusion power systems, such as Robert Bussard of the polywell and Eric Lerner and his dense plasma focus, argue that all the research money is being diverted into this one project to the detriment of other promising technologies. As the situation stands, there is insufficient government funding to pursue every fusion project, and the worldwide majority of plasma physicists believes that ITER is the best prospect.

Construction of the ITER complex in France began in 2008, with installation of the fusion chamber to begin in 2011 and to be completed in 2018. Operation of the reactor is expected to continue for 20 years, after which enough should be learned so that a power reactor can be built. Constant upgrades are being made to the design of the reactor even before construction is started, and it is not unreasonable to expect some physical modifications before it is completed. ITER is an experiment in which unknown properties of a fusion reactor operating beyond the ignition temperature are to be found, and design changes are part of this process. Unfortunately, dizzying cost overruns are also part of this process. It is hoped that the consortium realizes this and can exercise patience.

THE NATIONAL IGNITION FACILITY

Magnetic confinement fusion may be confirmed as the preferred method of fusion for power production by the ITER experiment if it is successful. If not, there is the National Ignition Facility (NIF) at Lawrence Livermore National Laboratory in Livermore, California. NIF is the largest, most powerful laser system in the world, and with it scientists hope to reach the ignition point for inertial confined fusion. If NIF is successful in causing most of an entire loading of deuterium-tritium to fuse with a single blast of laser light, further work in the next 40 years could develop a laser-fusion power plant.

The NIF makes use of lessons learned in all previous laser-fusion experiments conducted over the past 40 years at Lawrence Livermore. With the last series of large laser experiments using the NOVA facility, it became obvious that a bigger machine would be necessary to achieve the goal of self-sustained fusion, or ignition. In the case of laser-fusion, the self-sustained phase is always a short-lived pulse, and there is no attempt to make it continuous. It is necessary to exhaust all the fuel in one shot or there is not enough energy released to justify the energy used to fire and amplify the lasers. For electrical power production, pulses will have to be repeated regularly, and this problem has yet to be addressed. First, it must be proven that a single pulse can reach the ignition point.

NIF was designed as a follow-on to the NOVA laser-fusion project in 1994. The projected budget was $1 billion with construction to be completed in 2002. This large expenditure was justified by tying the fusion-power experiment to the Stockpile Stewardship and Management Program (SSMP), which was intended to enable nuclear weapons testing without setting off full-sized bombs underground. Causing a tiny, frozen pellet of deuterium-tritium, the size of a BB, to fully fuse by laser light was the same as setting off a 10-megaton thermonuclear warhead, only in miniature. Much the same data could be collected from the laser-fusion shot as could be taken in an expensive, diplomatically unattractive weapons test. In 2002, uranium and plutonium fission were added to the NIF test schedule, subjecting very small samples to the same stress as would be encountered in a full-sized bomb explosion. These tests had nothing to do with fusion power, but they helped justify the expanding budget.

In 1997, the funding agency, the U.S. Department of Energy, pushed the completion date to 2004 and added $100 million to the budget. In 1998, the project needed another $100 million. The first flash-lamp testing was in October 1998. The project team grew alarmed when the capacitor bank

The NOVA laser target chamber (*Lawrence Livermore National Laboratory*)

for the pulsed power conditioning modules exploded, sending debris scattering through the mechanically sensitive optical equipment in the laser gallery. The building, which is the size of three football fields, had to be redesigned to protect the equipment against the violent, unexpected disassembly of adjacent units. Another $350 million was required, and the completion date was pushed to 2006. An examination of the progress of NIF in 2000 set the new budget at $3.3 billion. A year later, a follow-up report reset the budget to $4.2 billion and predicted that the facility might be completed sometime in 2008. On January 26, 2009, the first major construction milestone was completed, and on February 26, 2009, the laser was test-fired into the target chamber.

The NIF is impressive in its size, power, and precision in landing multiple laser beams on a very small target simultaneously, with the timing down to trillionths of a second. When used to full capacity, the NIF will deliver a 500-trillion watt flash of ultraviolet light into a hohlraum the size of a pencil eraser. The flash starts in a single, infrared laser. The short pulse of light from this laser is split into 48 beams and sent through preamplifiers. In this step, the few billionths of a joule from the laser is increased to six joules.

The beams are then directed into the main amplifiers, which are pumped by 7,600 xenon flash lamps. These lamps are similar to but much larger than the electronic flash used on digital cameras. The amplifier components in the beam-lines are about the size of small cars, and the length of one line from the original laser to the target chamber is 1,000 feet (300m). At this point, the beams have been amplified to 4 million joules. As a last step, the beams are converted from infrared to ultraviolet light using thin sheets cut from a single crystal of potassium dihydrogen phosphate. This frequency conversion is not very efficient, and the beams are cut down to 1.8 million joules. Spatial filters, or refractor telescopes, are used to refocus each beam down to a smaller spot after it is subjected to amplification.

At the final intersection of these 48 beams is the target chamber, weighing 287,000 pounds (130,000 kg). The light pulse is focused on the center of the spherical chamber, hitting the hohlraum from all sides. The gold sides of the hohlraum vaporize, and a shock wave of newly generated X-rays hits the frozen deuterium-tritium mixture. The outer layer of frozen gas flashes off under the X-ray bombardment, causing shock waves to both radiate out and implode into the center of the target pellet. If all goes according to plan, the middle of the target under the pressure of the imploding shock wave will fuse and continue to fuse as the heat of the reaction leads to higher temperature and pressure. The only thing holding the pellet together will be its inertia, which will be effective for billionths of a second, and in this short time most of the deuterium and tritium must fuse. Optimum fusion will create a miniature nuclear explosion, the equivalent of 0.000012 kilotons of TNT, or 45 million joules of fusion energy. This is far short of the 400 million joules needed to charge the capacitors for the beam amplifiers. A break-even fusion event will be the next problem to be solved, for a bigger laser. A repetition rate of one shot per day may be possible, as the amplifiers must be allowed to cool down between firings.

Seen from above, each of NIF's two identical laser bays has two clusters of 48 beamlines—one on either side of the utility spine running down the middle of the bay. *(Lawrence Livermore National Laboratory)*

The NIF accomplishments so far include winning the 2009 Digital Video Awards International Competition in eight different categories, including visual effects, scriptwriting, 3D animation, and art direction for its production of "The Power of Light." If it is as successful in its quest for laser-induced ignition, then power production by inertial confinement could be here in the next century.

THE DEMO POWER PLANT: ULTIMATE PROOF OF FUSION POWER GENERATION

If the ITER experiment is at all successful, then an even larger tokamak fusion reactor must be built to make use of the findings and actually demonstrate the generation of electricity by fusion power. It has been under tentative development since 2004, named the Demonstration Power Plant, or simply DEMO. The schedule for this facility is preliminary and subject to change, depending mainly on developments at ITER. If all goes according to plan, the conceptual design for DEMO will be complete by 2017, with an engineering design by 2024. Construction will proceed from 2024 to 2033, after which the reactor will be run for a five-year test. Beyond 2040, an expanded and updated plant will be operated. This schedule assumes that ITER is successful in meeting all its goals on time.

DEMO will deliver 2.3 billion watts of electricity, making it as powerful as a fission power plant having two running reactors. The machine must be of unprecedented size, having a fusion chamber 38 feet (11.6 m) in outer diameter, or 15 percent larger than ITER with 30 percent more plasma density. It must run continuously, whereas ITER only promises to try to run for 1,000 seconds. ITER plans to make 10 times more power than is required to run it. DEMO has to produce 25 times more power than it uses to be economically feasible. These are tall goals, but these specifications are necessary if fusion power is to become reality.

The fusion chamber lining, made of ceramic tiles, will be cooled by liquid lithium running through embedded pipes. Lithium-6 in the coolant will absorb neutrons and convert into tritium, which will be scrubbed from the liquid and recycled back into the deuterium-tritium fusion process. Helium, which is also a product of the tritium breeding, will be removed for use in the superconducting magnet coolers. To achieve the self-sustaining fusion, the inside of the fusion chamber must be kept at more than 100 million degrees. If the superhot dense plasma should accidentally touch the walls of the fusion chamber, it will burn right through.

All hope is not invested in ITER. If ITER should fall short of expectations, there is another, alternate large-plasma experiment being built in Greifswald, Germany, for completion in 2015. It is the Wendelstein 7-X, a stellarator of advanced design and a backup if ITER fails to reach ignition. Its goals are to operate continuously for 30 minutes with a temperature of 100 million degrees in the twisted-torus fusion chamber. The concept of continuous operation is an essential property of a power plant, and this stellarator promises to outperform the ITER in this respect. A collection of 50 nonplanar magnet coils will adjust the plasma ring in a precise, delicate fashion, unlike the brute force approach of a tokamak. Most of the money is going into the ITER tokamak, but the stellarator concept is not dead. The DEMO design is not so committed that it could not be turned to this alternate avenue.

The site, funding mechanism, and budget for DEMO have yet to be determined. There is an enormous engineering load to be handled. If ITER and then DEMO are successful projects, then in the last quarter of this century mankind will enter the age of fusion. Civilization will be powered with an inexhaustible, environmentally friendly, and universally available resource, hydrogen fusion.

Conclusion

After 60 years of research, development, and experimentation in the United States, Europe, China, Japan, India, Australia, and South America, following several divergent ideas and theoretical concepts, with the expenditure of more than $100 billion, there is not a fusion reactor that can claim to have produced more energy than is required to initiate the reaction. Although an enormous amount of work has gone into it, a machine that produces usable power by fusion has yet to be constructed.

The task of building a confined environment on Earth that far exceeds the temperature and pressure at the center of the Sun has proven more difficult than was initially thought. As a result of a lack of complete success in some early experiments, ever larger, more precisely defined plasma environments have been built. The complexity, the exotic materials, and the fragility of the latest fusion systems are overwhelming.

A comparison must be made to the companion nuclear power experiment at the other end of the curve of binding energy, the fission of uranium. The first nuclear fission reactor that produced power on an industrial scale was the B Reactor at the Hanford Works in Richland, Washington. It ran without incident for 24 years, 24 hours a day. When first started, it developed 250 million watts of thermal power, but by the time it was finally shut down it had been upgraded, and this single reactor output 2.21 billion watts of continuous heat. Its design was based on three months of research and experimentation on the CP-1 reactor at the University of Chicago and construction of a small prototype, the X-10 reactor at Oak Ridge, Tennessee.

This machine was built using all the industrial capabilities of 1944. Compared to what is now available in the 21st century, the equipment and techniques available back then were somewhat rudimentary. There was no solid-state electronics industry in 1944, no integrated circuits, and no computers. Graphite without boron contaminants was an exotic material. Design calculations, building tolerances, and instrument readouts were made to three or fewer significant figures. Thermal and biological shields around the reactor were made of cast iron and masonite.

Many technical advances have occurred since then, but nuclear fission reactors are still built using standard industrial tolerances and techniques.

Nothing particularly special is required to build a nuclear plant, with the exception of enhanced safety and security qualifications. The major efforts are in concrete forming for the building, steel forging for the reactor vessel, and pipe-fitting, which is still the same skilled craft it was in 1944. Mechanical fit tolerances of metal components need only follow the codes and standards of the American Society of Mechanical Engineers. Further building standards and practices are specified by the Institute of Electrical and Electronic Engineers and the U.S. Nuclear Regulatory Commission. While these building and operating rules are conservative, none is so restrictive as to make nuclear power too expensive to possibly compete in the open market of energy production. Uranium fuel processing and forming techniques in use today are basically the same developed in 1944.

Sticking closely to standard, familiar industrial practices and using craft labor from other industries are attractive to the electrical power industry. The public-owned utilities have thus made a cautious, gradual transition to nuclear fission power from more traditional, environmentally hazardous power generation schemes. The bottom line for these companies is to make money by selling electricity, by whatever means are allowed, and nuclear fission has made some inroads into this very practical, capitalistic endeavor.

If it is to succeed as a power production system, fusion must drop into this same environment. It may not be enough that fusion power will not make radioactive waste. It has to not make radioactive waste while making electrical power cheaper than fission. The bottom line of all the profits minus all the expenses is a complicated piece of accounting. It involves everything from the fuel production to the teardown of a worn-out power plant, and basically none of these expenses is yet known for fusion power. Fuel production for fusion, for example, will require the extremely expensive isotope separation of deuterium from water, the mining of lithium, the isotope separation of lithium-6 from natural lithium, and the conversion of lithium-6 to tritium.

Techniques, practices, tolerances, and materials for making experimental fusion reactors are far beyond standard industrial methods. Helium-3, for example, must be liquefied for use in niobium-tin superconducting magnets, and helium-3 is not an industrial material. A neodymium-glass laser amplifier of the size used in laser fusion is not a manufactured item. Industry may step up to these new concepts if the need arises, but fusion power systems cannot be manufactured in a practical way at this time, as were fission reactors in 1944.

The danger of current fusion power research is that it could be successful, breaking through to the ignition point or to the break-even level. As expertise has advanced over the past decades and experiments have improved, future triumph is possible. With at least one such success, more money and effort will probably be poured into fusion research, and eventually a workable reactor could be designed and prototyped. Judging from the trajectory of fusion developments, such a machine will be too big, too complicated, and too costly to be competitive in the public-owned electrical generation industry. The utilities are able, barely, to endure fission power, with all its rudimentary, familiar industrial qualities. A fusion machine that requires liquid helium-3, liquid lithium-6 coolant, and a combustion chamber lined with diamonds is far beyond what the current power industry can tolerate.

These concerns may be groundless. The laser fusion device currently envisioned would have to make one precise blast of light per minute, efficiently fusing its entire fuel-pellet, to create usable power, and there is no workable plan for a mega-joule laser that can manage this duty cycle. One shot every five hours may be possible in the future if a cooling system can be devised. Nor is there a plan for a mechanism to remove power from the fusion chamber and direct it to a turbo-generator. Such complexities could stop the laser fusion effort in its tracks, even if a break-even ignition can be achieved.

Magnetic confinement fusion is farther along, and a fully operational tokamak-based power plant is now on the drawing screens. Complexities such as power removal and tritium conversion are incorporated into the design. However, plasma is stubbornly difficult to confine for a practical interval. Once it is so confined, the problems of wear and tear on the torus liner and all auxiliary equipment will come to the forefront, and another round of thorny engineering problems will have to be addressed. It will take time.

In the future, when fusion power plants have become practical to build and switch into the power grid, the concept of the publicly owned utility may have evolved and progressed to the point where fusion is a perfect fit. There is much work ahead.

1926   Fritz Paneth and Kurt Peters fuse hydrogen into helium, using palladium as a cold-fusion medium. After publishing the discovery in *Nature,* they find that their results were in error and retract the findings.

1927   Georges Lemaître, a Belgian Roman Catholic priest, proposes the big bang theory of the origin of the universe.

1929   Fritz Houtermans and Robert Atkinson theorize that hydrogen can be fused into helium and that this reaction will release net energy.

1932   Mark Oliphant at the Cavendish Laboratory discovers tritium.

Mark Oliphant fuses together deuterium and hydrogen using a particle accelerator, making helium-3.

1938   Eastman Jacobs and Arthur Kantowitz build a magnetic-pinch torus fusion reactor at the Langley Memorial Aeronautical Laboratory.

1939   Hans Bethe formulates a theoretical model for fusion reactions in stars.

1941   Enrico Fermi proposes the idea of a hydrogen fusion bomb to Edward Teller.

Teller becomes obsessed with the thermonuclear weapon concept and neglects his work on the fission-powered atomic bomb.

1951   Ronald Richter announces successful controlled nuclear fusion for power generation, developed at a secret lab on Huemul Island.

Lyman Spitzer starts Project Matterhorn at Princeton University, developing the stellarator fusion reactor.

James Tuck starts Project Sherwood at Los Alamos National Laboratories to develop the perhapsatron plasma-pinch reactor.

In the Soviet Union, Igor Tamm and Andrei Sakharov begin conceptual design of the tokamak reactor.

In Great Britain, theoretical work begins on the Zero-Energy Thermonuclear Assembly.

Edward Teller and Stanislaw Ulam formulate the Teller-Ulam thermonuclear weapon design.

1952 The Ivy Mike hydrogen bomb test is successful, yielding an explosion equivalent to 10 megatons of TNT and evaporating an island in the Pacific Ocean.

1953 Fusion reactors in the United States and the Soviet Union make neutrons, but not by fusion.

1954 The ZETA fusion device in Great Britain is operational.

1956 At the Kurchatov Institute in Moscow, research begins with tokamak T-1.

1958 On **January 28,** British newspapers report breakthrough success in the ZETA Project at the Atomic Energy Research Establishment.

In **May,** British fusion researchers find that there was actually no fusion occurring in their ZETA device.

On **September 1–13,** Scientists meet together at the International Atoms for Peace Conference in Geneva, Switzerland, and secrecy is dropped for the first time.

1961 On **October 30,** Soviet Union detonates the world's most powerful fusion weapon, the Tsar Bomba. The explosion is the equivalent of 50 million tons of TNT.

1964 Arno Penzias and Robert Wilson interpret measured microwave noise as a signature of the big bang.

A Z-pinch fusion reactor is demonstrated to the public at the New York World's Fair.

1967 The Farnsworth-Hirsch Fusor is demonstrated, making neutrons in a fusion reaction on a tabletop.

1968 The T-3 Soviet tokamak produces temperatures 10 times greater than all other plasma machines.

1972 Inertial containment research begins at the Lawrence Livermore National Laboratory, with development of the "Long Path" neodymium-glass laser.

1974 Construction begins on the Janus twin-laser inertial containment experiment at Lawrence Livermore.

1975	A new laser for fusion research, the Cyclops, is tested at Lawrence Livermore.
1977	The 20-beam Shiva laser, the size of a football field, is built at Lawrence Livermore for fusion experiments.
1978	The Joint European Torus project is started at an ex-RAF airfield near Oxford, England.
1982	Work begins on the Tore Supra tokamak with superconducting magnets at Cadarache, France.
1983	On **June 25,** Joint European Torus is started.
1984	In **December,** the 10-beam NOVA laser is fired at a droplet of deuterium, attempting to make it fuse.
1985	The Japanese JT-60 torus is completed and begins experiments with simple hydrogen.
1988	The Soviet tokamak T-15 with superconducting magnets is completed.
	Conceptual designs for the International Thermonuclear Experimental Reactor begin.
	In **April,** Tore Supra superconducting tokamak begins operation in France.
1989	On **March 23,** Stanley Pons and Martin Fleischmann at the University of Utah announce successful energy production by fusion in a glass of heavy water. The claim was subsequently invalidated.
1992	Engineering design activity begins for the ITER reactor, with EURATOM, Japan, Russian, and the United States participating.
1993	In **December,** the Tokamak Fusion Test Reactor at Princeton University produces 10 megawatts of power from a controlled fusion reaction using a deuterium-tritium reaction.
1996	Tore Supra achieves a record plasma duration of two minutes,
1997	The JET reactor in the United Kingdom produces 16 megawatts of power by fusion.
	Construction begins on the National Ignition Facility at Lawrence Livermore National Laboratory.

1999   United States withdraws from the ITER project.

2002   Small-scale fusion is discovered in collapsing bubbles in a tabletop apparatus at the Oak Ridge National Laboratory.

2003   The United States rejoins the ITER Project. China and the Republic of Korea have joined, while Canada joined and then dropped out.

            On **April 7,** Sandia National Laboratories achieves deuterium fusion with the Z machine.

2005   Cadarache, France, is chosen as the site of the ITER reactor.

            National Ignition Facility at Lawrence Livermore fires a small bundle of eight lasers, achieving the record for the highest energy pulse from a laser, at 152,800 joules of infrared energy.

2006   On **September 28,** China starts up the Experimental Advanced Superconducting Tokamak in the city of Hefei.

2009   Construction of the National Ignition Facility at Lawrence Livermore is complete.

2011   On **July 1,** The National Ignition facility successfully fires a 1.42 megajoule burst of laser light into an inert test target made of silicon and plastic

2015   Work begins on contruction of the ITER fusion reactor, with completion in 2018 and a start-up in 2019.

Glossary

absolute zero   theoretical temperature at which entropy, or the movement of molecules, would reach its minimum value

ASDEX (Axially Symmetric Divertor Experiment)   fusion reactor at the Max Planck Institut für Plasmaphysik in Garching, Germany; built in 1991, the largest fusion device in Germany

Atomic Energy Research Establishment (AERE)   main center for atomic energy research and development in the United Kingdom near Harwell, Oxfordshire, from the 1940s to the 1990s

big bang   cosmological model of the initial conditions and subsequent development of the universe, believed to have occurred about 13.5 billion years ago

CNO cycle   the process by which simple hydrogen fuses to form helium-4, using nitrogen-13, carbon-13, oxygen-15, and nitrogen-15 as intermediate stages of reaction

cosmic microwave background radiation   radio waves at 160.2 gigahertz that fill the universe; thought to be an artifact of the big bang, in which the known universe was created

deuterium   lighter of two possible isotopes of hydrogen, having one proton and one neutron

EAST (Experimental Advanced Superconducting Tokamak)   fusion reactor at the Institute of Plasma Physics in Hefei, China

fusor   vacuum tube that creates nuclear fusion by inertial electrostatic confinement, invented by Philo T. Farnsworth

hohlraum   small cylinder made of thin gold used to hold a sample of deuterium-tritium fuel to be fused in an inertial confinement device; replicates the conditions in a hydrogen bomb by producing a shock wave of X-rays to compress the fuel

Huemul Project   secret fusion power project sponsored by Argentina, beginning in 1948; success at achieving hydrogen fusion was announced in 1951 but later found to have been premature

inertial confinement fusion (ICF) method of generating fusion power, in which the exploding fuel is kept together by its own inertia

ITER (International Thermonuclear Experimental Reactor) ongoing fusion engineering project by the European Union, Japan, Russia, China, Korea, and the United States

Joint European Torus (JET) fusion reactor built at the Culham Science Centre in the United Kingdom; holds the record of 16 megawatts for power produced by nuclear fusion

Large Helical Device largest superconducting stellarator in the world, located in Toki, Gifu, Japan

laser light amplification by stimulated emission of radiation, a source of light at a single frequency, with all light waves vibrating in phase, or coherence

Laser Mégajoule (LMJ) an experimental inertial confinement fusion device being built near Bordeaux, France; will be as powerful as NIF at Lawrence Livermore and used to refine fusion calculations for the French nuclear weapons industry

National Compact Stellarator Experiment (NCSX) plasma confinement experiment conducted at Princeton Plasma Physics Laboratory; NCSX project was cancelled by the Department of Energy in 2009

National Ignition Facility (NIF) laser-based inertial confinement research device at Lawrence Livermore National Laboratory in Livermore, California, completed in 2009; expected to achieve fusion-ignition of deuterium-tritium fuel

neutral beam injection method of i ncreasing the temperature of plasma by introducing hydrogen atoms accelerated to high speed

nuclear binding energy average binding energy per nucleon (neutron or proton) in the nucleus of an atom

nuclear fusion process by which multiple atomic nuclei join together to form a single heavier nucleus

nucleosynthesis stellar process of creating new atomic nuclei from preexisting protons and neutrons

PACER Project project at the Los Alamos National Laboratory in the 1970s to produce fusion power by the explosion of hydrogen bombs underground

Perhapsatron   Z-pinch fusion reactor built by James L. Tuck at Los Alamos National Laboratories under Project Sherwood

Princeton Plasma Physics Laboratory   U.S. Department of Energy national laboratory for plasma physics and nuclear fusion science located in Plainsboro Township, New Jersey

proton-proton chain reaction   most basic fusion reaction, as occurs in stars, by which hydrogen nuclei fuse to form helium

SN1987A   supernova on the outskirts of the Tarantula Nebula in the Large Magellanic Cloud; light from SN1987A reached Earth on February 23, 1987

START (Small Tight Aspect Ratio Tokamak)   fusion reactor at the Culham Science Centre in the United Kingdom; built in 1991 and in 1998 was given to ENEA research laboratory in Frascati, Italy

stellarator   device used to confine a hot plasma with an external magnetic field for the purpose of sustained fusion

superconducting magnet   electromagnet made from coils of wire having zero electrical resistance

supernova   rare, unusually violent stellar explosion; light from a supernova can outshine an entire galaxy of stars

Teller-Ulam device   successful design for a megaton-range thermonuclear weapon, using a plutonium-based fission bomb and an adjacent cylinder of lithium-6-deuteride within a uranium-238 tamper

tokamak   fusion reactor using both a Z-pinch and two external magnetic fields to confine and compress plasma in a toroidal, or donut-shaped, chamber

Tore Supra   one of the largest tokamak fusion reactors built using superconducting magnets, at the nuclear research center of Cadarache, France

tritium   the heaviest isotope of hydrogen, consisting of one proton and two neutrons

United Kingdom Atomic Energy Authority (UKAEA)   established on July 19, 1954, to make pioneering advancements in nuclear power

Zero Energy Thermonuclear Assembly (ZETA)   Z-pinch fusion reactor built in the United Kingdom in 1954; in 1958, ZETA research team erroneously reported successful hydrogen fusion

Z-pinch   type of inertial plasma containment in which compression is created by an electrical current through the plasma

Further Resources

BOOKS

Close, Frank. *Too Hot to Handle: The Race for Cold Fusion.* Princeton, N.J.: Princeton University Press, 1991. An excellent description of the frantic efforts to validate cold fusion findings in 1989.

Harms, A. A., K. F. Schoepf, G. H. Miley, and D. R. Kingdon. *Principles of Fusion Energy: An Introduction to Fusion Energy for Students of Science and Engineering.* Singapore: World Scientific Publishing, 2000. A graduate-level discussion of the technical details of fusion reactors. Requires knowledge of calculus, linear algebra, and operational mathematics.

Herman, Robin. *Fusion: The Search for Endless Energy.* New York: Cambridge University Press, 1990. Not up-to-date, but a good introduction to the history of fusion power research with understandable technical descriptions.

Huizenga, John R. *Cold Fusion: The Scientific Fiasco of the Century.* New York: Oxford University Press, 1993. The most complete evaluation of the science of cold fusion and its downfall.

Mariscotti, Mario. *Secreto Atomico de Huemul.* Buenos Aires, Argentina: Sudamerica, 1987. Written in Spanish, the only published description of the Huemul Project in Argentina. Includes many rare photographs of Ronald Richter and his laboratory setup.

Peat, F. David. *Cold Fusion: The Making of a Scientific Controversy.* Chicago: Contemporary Books, 1989. One of the first histories of cold fusion to be published, concentrating on Stanley Pons and Martin Fleischmann at the University of Utah.

Seife, Charles. *Sun in a Bottle: The Strange History of Fusion and the Science of Wishful Thinking.* New York: Viking, 2008. A very readable though critical introduction to the history of fusion power research.

Taubes, Gary. *Bad Science: The Short Life and Weird Times of Cold Fusion.* New York: Random House, 1993. A thorough look at how science works by an examination of the cold fusion controversy in 1989.

WEB SITES

Further depth on many topics covered in this book has become available on the World Wide Web. Biographies of famous scientists are available, and a deeper probing into the lives and the careers of notable physicists and chemists can be worth pursuing. Always look for cross-referencing links in a Web site. A double click on a highlighted word can take you to even further depth, giving details on interesting subtopics. Also available are detailed Web sites for any government facility or agency, some giving access to archives and histories.

Bubble Fusion is referenced on this Internet site, giving a summary of the experiments and a good collection of references to other sites and articles. Available online. URL: http://inventors.about.com/od/tstartinventors/a/ Rusi_Taleyarkha.htm. Accessed June 20, 2011.

Cold fusion, now called "lattice-assisted nuclear reactions," is not completely dead. Read all about it in the Cold Fusion Times, at this Web site. Available online. URL: http://world.std.com/~mica/cft.html. Accessed June 20, 2011.

DEMO is the fusion power plant that is to be built using the research from the ITER project. This paper from the *Journal of Research from the National Institute of Standards and Technology* details the engineering challenges that this proposed project will face. Available online. URL: http://math. nist.gov/~GMcFadden/GaMc09.pdf. Accessed June 20, 2011.

Dense Plasma Focus research is alive and well and detailed on the Focus Fusion Society Web site. Included are active forums and milestone updates. Available online. URL: http://www.focusfusion.org. Accessed June 20, 2011.

ITER is the large tokamak project underway in France. The official page for this international undertaking includes a video, slide show, cutaway drawings of the reactor, news updates, and even a job application. Available online. URL: http://www.iter.org. Accessed June 20, 2011.

LANL Research Library is a vast collection of eBooks, databases, and news concerning all issues of nuclear science, nuclear technology, the history of nuclear topics, and current research and development. Students of nuclear science will find it an invaluable resource. The library search can find books, journals, patents, reports, videos, audiotapes, and recommended Web sites. Available online. URL: http://library.lanl.gov. Accessed June 20, 2011.

Los Alamos National Laboratory is an online science and technology magazine with excellent access and interest for educators and students. It includes articles concerning current work at the Los Alamos National Laboratory, with sections concerning the environment, business and technical transfers, and postdoctoral studies. A link to the Bradbury Science Museum is included in its features. Available online. URL: http://www.lanl.gov/. Accessed June 20, 2011.

MIGMA fusion by Bogdon Maglich is covered at this site. Features include references to other information sites, reprints of articles describing the MIGMA reactor, and even two articles by Maglich himself. Available online. URL: http://www.rexresearch.com/maglich/maglich.htm. Accessed June 20, 2011.

Muon Catalyzed Fusion is explained in this PDF-format slide show presentation by Frederick Manley on March 8, 2007. Available online. URL: https://wiki.engr.illinois.edu/download/attachments/19302097/Manley+-+Muon+catalyzed+fusion.pdf?version=1&modificationDate=1232817345720. Accessed June 20, 2011.

National Ignition Facility, "Bringing Star Power to Earth," is the official Web site for the world's largest and most powerful laser. This site includes many stunning photos of the NIF, as well as animated presentations and up-to-date news on the progress of the fusion power experiments. Available online. URL: https://lasers.llnl.gov. Accessed June 20, 2011.

Polywell fusion is covered in this compendium of other sites and references dealing with initial and continuing research. Available online. URL: http://www.strout.net/info/science/polywell/index.html. Accessed June 20, 2011.

Project PACER never reached a full test of its capabilities as a power source but is still discussed as a possible fusion reactor. The Plowshare Program in general, of which PACER was a part, is discussed at this Web site. Available online. URL: http://www.world-nuclear.org/info/PNE-Peaceful-Nuclear-Explosions-inf126.html. Accessed June 20, 2011.

Tokamaks all over the world are listed in this compendium of tokamak fusion research. Specifications of all 209 tokamak machines are given, as well as photographs, a list of discoveries made using tokamaks, and a special section covering spherical tokamaks. Available online. URL: http://www.toodlepip.com/tokamak. Accessed June 20, 2011.

Stellarator Wendelstein 7-X information is available directly from the source, the Max Planck Institute for Plasma Physics. Available online. URL: http://www.ipp.mpg.de/ippcms/eng/pr/forschung/w7x/index.html. Accessed June 20, 2011.

United States Department of Energy has a rich, complete site spanning the wide interests of this government agency, from energy science and technology to national security of energy sources. The Department of Energy owns, secures, and manages all the nuclear weapons in the military inventory. Although it concerns all sources of energy, nuclear power is a large component of its mission. This Web site is particularly accessible to students and educators. Available online. URL: http://www.doe.gov. Accessed June 20, 2011.

Italic page numbers indicate illustrations.